DON ROBINS

# The I$^2$C BUS

# The I²C BUS

## From Theory to Practice

**Dominique Paret**
*Philips Semiconductors,*
*Paris, France*

*with*

**Carl Fenger**
*Philips Semiconductors,*
*Zürich, Switzerland*

JOHN WILEY & SONS
Chichester • New York • Weinheim • Brisbane • Singapore • Toronto

Original French edition published as Le bus I²C
Copyright © 1994 by Dunod, Paris, France

Copyright © 1997 by John Wiley & Sons Ltd.
Baffins Lane, Chichester,
West Sussex PO19 1UD, England

National      01243 779777
International  (+44) 1243 779777

e-mail (for orders and customer service enquiries): cs-books@wiley.co.uk

Visit our Home Page on http://www.wiley.co.uk
or
http://www/wiley/com

Reprinted January 2000

*Other Wiley Editorial Offices*

John Wiley & Sons, Inc., 605 Third Avenue,
New York, NY 10158-0012, USA

VCH Verlagsgesellchaft mbH, Pappelallee 3,
D-69469 Weinheim, Germany

Jacaranda Wiley Ltd, 33 Park Road, Milton,
Queensland 4064, Australia

John Wiley & Sons (Asia) Pte Ltd, 2 Clementi Loop #02-01,
Jin Xing Distripark, Singapore 0512

John Wiley & Sons (Canada) Ltd, 22 Worcester Road,
Rexdale, Ontario M9W 1L1, Canada

British Library Cataloguing in Publication Data

A catalogue record for this book is available from the British Library

ISBN 0 471 96268 6

Typeset in 10½/13pt Sabon by Aarontype, Bristol
Printed and bound in Great Britain by Bookcraft (Bath) Ltd
This book is printed on acid-free paper responsibly manufactured from sustainable forestation,
for which at least two trees are planted for each one used for paper production.

# Contents

# Preface

Communication networks are arousing a great deal of interest at this time, particularly when they are used for so-called 'local' applications. For obvious reasons of economy, those based on serial buses are in the forefront at present and, particularly, the I²C bus, which we will discuss extensively throughout this work. Having personally participated in its development, from its inception to its most recent innovations, early in 1996, we are in an ideal position to help you know and understand this first-rate bus that everyone is talking about and using, often without being consciously aware.

The purpose of this work is, therefore, to describe this bus, as completely and precisely as possible, its secret life and its applications, without losing sight of the fact that, whatever the intellectual level of the subject matter, it will be easier to assimilate with a touch of humor, in a light and amusing style! If we accomplish all that, we will have risen to the challenge and accomplished our objective.

The presentation of the subject, and our approach to it, have been purposely designed so that everyone can find his place and adapt the level of his own knowledge to the material presented.

We have had technical students in mind, who will have to prepare and present prototypes in a matter of weeks, as well as students at engineering schools with more ambitious industrial 'projects'. As for professionals, in high level research and development laboratories, who are used to reading all sorts of electronic journals, they will very quickly recognize that part of the text that we have especially slipped in for them.

# Foreword

When Philips first invented the two-wire $I^2C$, or 'Inter-Integrated Circuit' bus in the late 1970's, it was never anticipated that 15 years later $I^2C$ would become a worldwide industry standard.

$I^2C$ was originally developed as a control bus for linking microcontroller and peripheral Ics for Philips consumer products. The elegant simplicity of a two wire bus combining both address and data bus functions was quickly adopted in such diverse applications as telecommunications (corded and wireless handsets in particular), automotive dashboard, PCs (as a diagnostics bus), energy management systems, test and measurement products, medical equipment, even in toys!

Spreading first throughout Europe, and then Asia, $I^2C$ has made major inroads in North America in the 90's where interest in $I^2C$ has risen dramatically. As soon as embedded systems designers realize the cost, space, and flexibility of this robust serial protocol, they're hooked on $I^2C$!

No longer a Philips-only concept, $I^2C$ has been licensed to over thirty manufacturers who now produce over 500 $I^2C$-bus compatible devices. $I^2C$-bus based system designs require no special licence, and the $I^2C$ protocol is easily implemented by virtually any microcontroller on the market.

With this book we hope to give the electronics student as well as the electronics systems designer a quick look into the world of $I^2C$, and an introduction to its principles and applications.

# Acknowledgements

A few friendly words of thanks to the many members of the various teams and laboratories of the Philips Semiconductors Company, mainly to M.M. Herman Schutte and A. P. M. Moelands Eindhoven, The Netherlands, M.M. Kevin Gardner and Peter Brown in Sunnyvale, California, USA for their collaboration and, also, a very special mention to Mr Carl Fenger International Product Marketing for $I^2C$ IC's in Zürich, Switzerland, who undertook corrections, modifications and adaptations of many details with respect to the publication of this English version of this $I^2C$ *bus* book.

I would also like to thank some French friends, Mesdames and Messrs. Marie-Laurence Cibot-Devaux, Blandine Delabre-Garnier and Jean-Pierre Billiard for their assistance in the realization of the various applications presented, as well as to Bernard Fighiera and Claude Ducros of the journal *Électronique Radio Plan* for their help and, finally, to my children Emmanuel and Fréderic for their patience, without which this work would certainly never have seen the light of day.

**Remark**
The cover illustration is based on a Philips Components document. Many of the drawings, with which this work has been illustrated, have been borrowed from Philips Semiconductors Technical Documentation.

# THE I$^2$C BUS

# 1 Waiting for the Bus

## Part 1

## General Remarks on Communication Buses

This book will be your guide through the world of microcontroller-managed serial buses, with special emphasis on the I²C Bus.

The following chapters present both theory and practice, exploring the controversies, trade-offs and compromises involved in choosing and designing the best possible control bus for your particular application. To make this book both useful and user-friendly, we have provided you with examples of modular solutions at various levels of complexity.

To give you the information needed to design effectively in a real-life environment, this book examines typical industrial and consumer bus applications. We have chosen these application frameworks to give you free rein to develop solutions to your own design problems. So, hang in there with us, and your reward will be a sound practical and theoretical understanding of microcontroller-managed control buses and I²C. Let's begin at the beginning.

In general, the building blocks of a micro-controlled system are:

- a microcontroller

- memories

- input/output interfaces (I/O).

Each of these components has its own specific characteristics. First, we'll examine the overall performance of the system, and then demonstrate how each element relates to the other building blocks of a microcontroller-managed bus.

# Performance and Characteristics Required of a System

Each of the following elements is a consideration in the planning and design phase of most system design projects:

- speed
- performance
- cost efficiency
- reliability
- conformity to recognized standards
- electromagnetic interference and propagation
- protection
- privacy
- conformity to regulations
- transmission medium.

Many of these requirements can be mutually contradictory. Your priorities depend on your application. After discussing each system component in detail, we will look at how each of these elements connects and interacts within the structure of a control bus.

# Speed

Many users choose to run their system at the highest possible clock frequency; possibly, because they lack precise project specifications. Why make your system run at light-speed when you can achieve the same objectives at a slower rate?

Of course, there are specific cases for which speed is essential, but not every application involves real time systems restricted to nanoseconds. It always makes sense to consider speed within the context of your own system, especially systems performing human interface. If you press a button and the desired result takes place in a thousandth of a second instead of a ten-thousandth of a second, who can tell? Resist the temptation to overdo it when you design for speed.

Although economics usually dominate your speed choices, cost savings are not the only reason for you to look at speed requirements realistically. The goal is to optimize the means you employ to realize the performance you require. If your application really calls for a 50 MHz, 32 bit parallel bus, use it. But if a 300 baud twisted pair bus does the job, always opt for simplicity.

By the same token, parallel processing obviously transmits more information faster than serial processing. This doesn't mean that parallel processing is necessarily the best solution for your particular design problem. The first rule of speed is: **Don't over-engineer your project.**

## Performance

In theory, you want a system capable of performing many functions of varying complexity with a high level of confidence. A control bus system that successfully connects various components has to manage timing and information flow. Your system's performance depends on a robust and dependable bus protocol.

Protocol resembles etiquette – the code of good manners that enables people to interact smoothly without interrupting each other and stepping on each other's toes. Like people in a well-regulated society, bus communication requires information to wait its turn in line, to connect at the proper time, to communicate without perturbances, glitches, and cross-talk.

There are common principles that dictate bus communications etiquette or protocol. Bus systems can have masters and slaves (the slaves sometimes change roles). Each component must know how to:

+ identify itself

+ query other components 'politely'

+ respond to its own name

+ know all the rules of polite society ("After you, Alphonse. No, after you, Gaston ...." "No, it's your turn," ... etc.)

+ follow and obey the established arbitration procedures in case of conflict.

Of course, the bus must handle the actual application information processing, as well as protocols! Do you need parallel processing to manage all these protocols efficiently? The following chapters demonstrate how serial buses can accomplish all these tasks successfully – and economically. Maximum speed or a simpler, albeit slower system with a minimal number of wires – that is the question!

## Cost efficiency

Now we have a real design problem! It's called cost efficiency – the bottom line. Every engineer has to deal with cost considerations, regardless of the ideal system you may theorize. So, let's get it over with now.

Financial and resource management is fundamental to commercial and industrial applications. These users are up against the competition; their product or service will sink or swim based on price as well as performance. No amount of performance will save a product that is priced higher than the market will bear. Cost considerations may involve the price of components, materials, manpower, investment, industrial earnings, and production line interruptions.

The following parameters all affect cost:

+ the intrinsic cost of components

+ the number of components

+ the semiconductor technology (MOS, bipolar, GaAs, ...)

+ the price and type of IC package

+ the silicon chip area of each IC

+ the number of pins

+ the relationship between all of these parameters

+ the wiring

+ the area of printed circuit the components occupy

+ the number of PCB interconnect layers (more layers = higher cost!)

+ the components' weight and sensitivity to vibrations

+ mounting time

+ testability of the component and its function

+ the availability of single or multiple sources

+ functional interchangeability of components (RAM or EEPROM)

♦ upward and downward system compatibility, permitting modularization and multi-level implementation

♦ the annual production volume of each component.

In the following chapters, we will show how most of these cost factors can be minimized with one solution!

# Reliability

How do we define reliability? What constitutes real reliability is a controversial concept.

Should a system be able to detect and correct all errors, or simply enable the rest of the system to continue operating without interruption? Should specific software measures be implemented to guard against the once-in-a-billion chance that an LED doesn't light when it should? In the end, the importance of each bit of information in a control system is determined by the real-world application. Bullet-proof redundancy and error correction may be critical in an aircraft control system, but not necessarily so in a telephone or in a VCR.

Reliability has become an engineering obsession when your system always needs another level of protection! Should we impose reasonable limits by quantifying the probability of perturbing phenomena or should we use ever-increasing levels of redundancy?

We can answer this question with other questions. Three questions for you to consider:

1. How {when?} does your transistor or I.C. 'go dead' – in short circuit or in open circuit... and will this crash the system?

2. Does the system need to be resistant to 'unauthorized' interference?

3. How important are economic considerations to your application?

Make reasonable requirements of your system, whether basic or ultra-sophisticated, and concentrate on finding specific practical solutions.

# Conformity with existing standards

The topic of standards generates as much controversy as reliability. There are standards and there are standards – on the local, national and international levels – as well as the *de facto* standards that industry uses.

Given the proliferation of standards, how can you adopt a standard that will be well received? There are reputable names, such as IEEE or VME, each authoritative in very specific domains. There are the *de facto* standards already in use because of economic or technical factors. A simple example of a *de facto* standard in the field of music is the serial transmission bus, MIDI (Musical Instrument Digital Interface). Why MIDI? This particular bus standard is neither technically better or worse than many others. It is popular because it is mass-produced and already in use by an ever-growing industrial base.

Other types of bus standards are already being used daily by millions of people and applications, including televisions, car radios, ATMs, public telephones, and automobile instrumentation. These applications may not be as sophisticated as the Cray One computer, but they are all successful examples of *de facto* bus communications in action. When developing your system, don't re-invent the wheel...conforming to an existing standard (whether 'real' or '*de facto*') will make life easier.

## Electromagnetic interference protection and propagation

Electromagnetic interference protection and propagation go together like a horse and carriage. Both together add up to an annoying engineering problem–cross-talk: unwanted induction of signals into neighbouring channels.

You have an entire arsenal of textbook solutions at your disposal, including differential inputs and outputs, shielded and coaxial cables, twisted pairs, and common mode rejection, all of which are responses to the problem of cross-talk.

Our recommendation is to treat the disease at its source. The best way is to build your EMI protection right into the chip, by designing a device with self-canceling EMI properties. The bottom line is whatever works best given your cost constraints. Use the best cross-talk solution you can afford that is mass-producible. Shielding, ground plains, decoupling capacitors, and careful PCB layout are the front line of interference control.

## Protection

Wouldn't it be satisfying to deliver a bus system that couldn't fail – a system that infallibly started up again even after you threw a handful of 3 mm screws into the works! Such a design feat may be possible in the best of all possible worlds, but in reality, the best we can do is to design an intelligent resistance to aggression into our systems. You can accomplish this by your choice of both hardware and software.

+ At the hardware level, choose an economical and robust bus configuration that satisifes the technical requirements of your application.

+ At the software level, design robust programs capable of dealing with disturbances during information exchange in an acceptable manner.

Although we can dream about designing a completely protected system, invulnerable to system failures or user error, realistically our goal is to find the best possible compromise between technical application requirements and economic realities.

# Security

More and more users are asking for another level of protection–protection against unauthorized access to highly confidential or competitive data. How can you design operating system software that prevents unauthorized access and achieves high levels of secrecy?

Fortunately, products are now available on the market that effectively protect the internal codes at the ROM or EPROM level, making the codes unreadable by various types of radiation, such as X-rays, or by disassembling the integrated circuit itself. Encryption/decryption techniques exist today that make data interception virtually impossible. There is a host of components well adapted to your application requirements. You can achieve an extremely high level of secrecy with protected ROM's, EPROM's, or OTP's.

# Following the rules – conforming to regulations

Unfortunately, ignorance of the law is no excuse! There are literally thousands of regulations that apply both to academic and commercial applications. You don't have to know all of them, but there are two groups of regulations you really can't ignore:

1. Safety regulations

2. Regulations involving electromagnetic interference (EMI).

For instance, for safety, there is a European regulation (originally called IEC 65) and a US regulation (UL94, etc) that deals with insulation voltages, the distances between conductors and leakage paths.

Why pay particular attention to safety regulations? Systems often need to communicate with other systems which are not connected to the same power supply. Due to this fact alone, it is essential to be concerned with safety regulations.

## Independence of the transmission medium

Bus systems are often designed for use with specific transmission mediums, such as differential pair or infrared. If you want to design a more versatile system that is compatible with a number of transmission media, you must factor in all their different requirements.

Each medium has its own specific properties, such as data flow rate, associated bandwidth and maximum transmission distance, parasitic capacitances, and sensitivity or immunity to perturbing signals. Each property involves special considerations.

The following list includes the principal transmission media in current use:

+ single wire (PCB traces)

+ shielded cables

+ coaxial cables

+ flat multi-wire cables

+ twisted pairs

+ sindex

+ optical fibers

+ infrared links

+ radio-frequency links

+ power line.

## Choosing between transmission mediums

How do you chose the optimal transmission medium for your application? Each medium presents both advantages and disadvantages. The solution is to find the best compromise for each application.

Finding the best compromise involves satisfying many parameters simultaneously. Engineering would be dull and monotonous without complications and design challenges. Ask yourself this question: which bus best satisfies the requirements of all the media listed above?

The answer, of course, is $I^2C$.

# Part 2

# Introduction to the I²C Bus

In the first section, we presented the principal considerations that govern your choice of bus system. Now, we'll describe how these parameters became a specific protocol. The following section breaks down each of the parameters and translates them into electronic terms.

## Information Transport Mode – Serial or in Parallel?

Consider these requirements:

1. what speed is required by the application?

2. bottom line costs, including:

   + the number of interconnections

   + the number of pins on the integrated circuits themselves.
     (Bear in mind that the fewer the pins required by the control bus the greater the number that remains for other functions. The fewer pins you use for a given function, the lower the cost!

   + connectors for interconnecting cards

   + the number of wires and the space occupied by connections between modules

   + the copper area occupied by electrical connections and the IC components themselves

   + reduced weight and volume, resulting in lower sensitivity to shock and vibration (EMI) and lower transportation costs.

These requirements call for the greatest possible simplification in printed circuit design, including:

+ fewest possible interconnections

+ reduced design time

+ a single-sided printed circuit board

+ adaptability to various types of cost-efficient transmission media, for instance IR, power lines, twisted pair.

As these issues become more important, the logical solution is a bus that contains the fewest possible interconnecting wires: a bus that efficiently transmits information serially rather than in parallel.


## Pros and cons of the serial bus

What are the advantages and disadvantages of choosing a serial bus?

The first possible objection is that each device connected to the bus must be capable of handling and processing the protocol of the serial transmissions. This means greater silicon area (either because of additional logic, or more software), and hence greater cost. A serial bus configuration only makes sense when the cost of the integrated bus interface is outweighed by the systems savings though smaller IC packaging, reduced pin-count, and smaller PCB.

As it turns out, the system economy achieved by a serial bus is substantially superior to the incremental cost imposed by the integration of the bus interface. The savings realized through reduced pin count, smaller packages and fewer PCB traces far outweighs the cost of a a slightly increased silicon area.

Bear in mind that there are many more slave functions than master functions in most systems, and that the slave interface is quite simple. For true cost-efficiency, the slave interfaces should therefore be as economical as possible, by excluding or using sparingly such features as signal regeneration, clocks, or oscillators.

Each of the circuits connected to the bus has specific functions to accomplish. The greater the extent of the useful function, the greater the ratio of the active portion of the component to its bus interface, making the apparent cost and area of the interface negligible. All of these considerations help us to select a protocol.

## What distance?

The faster you jog, the more out of breath you become and the shorter the distance you cover!

When we discuss communication speed, we also have to consider the associated distance that the bus can reasonably cover. There are two schools of thought. Either the communication is local, serving only components in the immediate environment or the system is designed to connect various distant systems that may be several or many miles apart. Some specifications call for simultaneous local and long-distance communications, adding more challenges to our design problem.

We need to choose a flexible bus that enables us to slow down the clock frequency and cover longer distances, if necessary. The speed-distance equation is one of the parameters we need to consider when characterizing a bus. Figure 1.1 shows a representative example of some of the major buses used to date.

The physical limits of this speed/distance relation are tied to the electrical capacitance between the various conductors, leading to modification of the shape of the transmitted signals (integration), delays due to the propagation time (tpd), causing temporal non-coincidence of signals (racing). In the case of bi-directional connections, problems arise due to bus conflicts caused by reflection of the signal or running into another incident signal.

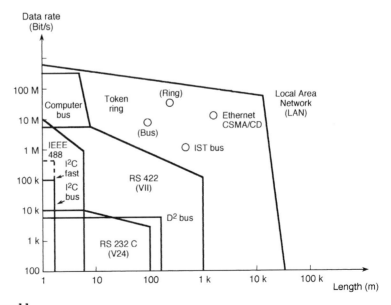

**Figure 1.1**

We can't decrease or totally cancel these physical capacitances. We can reduce the RC time constants by reducing the apparent values of the resistance seen at the line input. We achieve this by buffering the bus so as to attack the bus at lower impedance. This method makes it relatively easy to employ greater line lengths, although the overall power consumption will inevitably increase.

## How many wires in a serial bus, which wires, and why?

### One, two, three

#### One:

One of the problems with a serial bus is that it must herd the signals as they arrive, and then separate the 'sheep' from the 'goats'.

You can do everything with a single wire bus, for example transmit and receive information at the same time by multiplexing signal frequencies. This implies creating frequencies and demodulators that use up space on the chip. Other encoding schemes require more circuitry to recover the clock from the data.

The cost efficiency of serial connections is obvious, but the cost of the interface is nevertheless part of the equation, and a one-wire implementation is expensive.

#### Two:

There are a number of two-wire buses on the market, genuine and fake.

Genuine two-wire buses really have two wires, such as telephone pairs or differential pairs. If you look closely and take the ground return into account, these buses have almost three wires. However, the two-wire bus only needs two physical conductors to connect.

Almost-real two-wire buses definitely need two wires plus a ground return in order to function. Loosely speaking, you can say that ITT InterMetal's IM bus and Philip's I²C are two-wire buses. These are specific manufacturer proprietary buses used because of their particular application fields or performance.

The RS232 interconnection exemplifies the "fake" two-wire bus. When you carefully study the specifications in the V24 notice of the CCITT, you can recognize the enormous differences that exist between the simple structure of a two-wire connection and the V24 structure. However, with luck you can often get the RS232 to work with only two wires.

### Three:

We can conclude that a three(+) wire bus, such as Motorola's Serial Peripheral Interface (SPI) bus looks a little bit like a parallel bus and can't be described accurately as a serial bus. There really is no better choice–accept no more than two!

## Which wires?

What kind of information can two wires carry? There must be a clock signal, addressing data, control data, data relevant for the application, and traffic control information.

The simplest and most reliable bus interface is synchronous, with the clock generated by a single source. This implies that one wire must be dedicated for the clock. However it does not imply that you know who will be the system clock-provider in any particular time slice, or that the timing must be done at constant frequency (In certain modes of serial asynchronous connection (RS232, for example), you must often declare the data rate initially to avoid losing time by having to hunt for it).

How about the second wires? Now we can try to pass information from one device to another after we declare to whom the information is addressed. We can then transmit the data.

It would be nice to have a third wire which could act as our 'bus policeman' to resolve bus collisions should two bus masters conflict. What if you could accomplish this without needing the 3rd wire? Read on...

## Now, what do we do?

This question deserves special attention because everything on our wish list will sooner or later appear in our protocol. First, let's compile as precise a list as possible of our requirements.

- Data transfer must be bi-directional from one element to another (both reading and writing)

- Each of the circuits must function as a transmitter or receiver, as needed.

- Each circuit can be a master or a slave, determined either by its function at a specific point in time.

- Each circuit must have its own unique identity.

- An acknowledgment procedure must exist that informs a transmitter that the correct destination device has received its message or instruction.

- It should be possible to connect multiple masters to the bus, each capable of taking control of the bus.

- The unit that imposes its clock on the bus is the declared master.

- The master defines the bus speed.

- The physical or electrical presence or absence of a circuit in no way perturbs the operation of the bus.

- A bus conflict management protocol must exist that is capable of managing multi-master systems without losing either time or information.

- The ranges of electrical levels must be compatible with different technologies (bipolar, MOS, etc.), as well as different logical systems within the same technology (TTL, I2L, etc.), so that users with different applications can simultaneously connect to the same bus.

- The choice of protocol and/or transmission format should be open to allow future extensions or codes without being penalized.

- The software should be easy to install and operate and capable of processing information rapidly.

- The software should be easily implemented on any microprocessor/microcontroller.

You think achieving all this simultaneously on two-wire is impossible?

## Back to Earth

Now you're aware of the major parameters to consider when implementing a system based on a serial bus. Such a system could easily require tens of man-years to develope. Fortunately, Philips designers have already considered all these parameters when they developed the I$^2$C (Inter-Integrated Circuit Bus)!

The following chapters describe the I$^2$C bus in more detail and use the I$^2$C as the basis for modular examples of applications development, implementation, and trouble-shooting.

# 2

# The I²C Bus

## A Brief History of the I²C Bus

The I²C bus was introduced by Philips at the beginning of the 1980s as a standardized bus interface to support a growing family of general purpose and application specific IC's used in mass produced consumer electronics.

Since its introduction, the I²C bus has been chosen as the system control bus used in millions of televisions, telephones, radio receivers, car radios and other consumer electronic products. Originally used for controlling frequency synthesizers, IR encoding/decoding, keyboard decoders, LED/LCD display, low-speed memory, etc., I²C is now employed in such diverse applications as medical systems, disk drives, PCs, energy management systems, mainframe diagnostics bus, security systems and even toys.

As time progresses, more and more state-of-the-art designs are finding increasing uses for this versatile serial bus. Hundreds of small and medium sized enterprises as well as giant multinationals have already discovered the advantage of the I²C bus, and fully exploit it in their designs.

High volume industries operating with tight profit margins were immediately attracted to I²C, such as the automobile industry, which has been using the I²C bus to control many of its vehicle equipment since its introduction. Modern telephone design, with extended memory, frequency synthesizers (for cordless/mobile), Realtime clock, tone generators, LED/LCD displays etc. have especially benefited from the cost savings, physical size, and modularity of the I²C bus.

As I²C has grown into a worldwide serial bus standard, Philips has licensed it to over 35 manufacturers allowing them to grow and diversify the range of I²C bus compatible devices. Today, I²C is no longer a Philips proprietary bus: it belongs to the industry!

# The I²C Markets

Table 2.1 shows an overview of possible applications for the component and systems market of the conventional I²C .

**Table 2.1**

| Appliances | TV | Radio | Audio | Telephony | Industrial | Household | Automotive |
|---|---|---|---|---|---|---|---|
| TV reception | • | | | | | | |
| Radio reception | | • | | | | | |
| Audio processing | • | • | • | • | • | • | • |
| Infra-red control | • | • | • | | • | • | • |
| LCD display control | • | • | • | • | • | • | • |
| LED display control | • | • | • | • | • | • | • |
| DTMF | | | | • | • | • | |
| I/O | • | • | • | • | • | • | • |
| A/D & D/A converters | • | • | • | (•) | • | • | • |
| Clocks/timers | • | • | • | • | • | • | • |
| RAM | • | • | • | • | • | • | • |
| EEPROM | • | • | • | • | • | • | • |
| 8 bit microprocessors | • | • | • | • | • | • | • |
| 16 bit microprocessors | • | | | | • | • | • |

I²C is a medium speed bus with an impressive list of features:

+ resistant to glitches and noise

+ supported by a large and diverse range of peripheral devices

+ a well-known robust protocol

+ a long track record in the field

+ a respectable communication distance

+ compatibility with a number of processors with integrated I²C port (micro 8,16/32 bits) in 8048, 80C51 or 6800 and 68xxx architectures

+ easily emulated in software by any microcontroller

+ available from a number of component manufacturers

...this is a list that makes any engineer sit up and take notice.

Digital Electronics Corporation (DEC) has already agreed with Philips to create a derivative of the I²C bus, called ACCESS.Bus, to connect to all environments. See Figure 2.1.

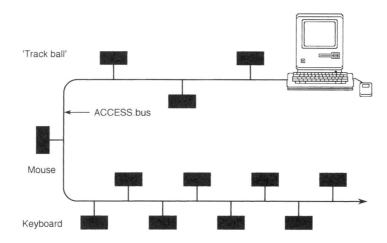

**Figure 2.1**

# The ACCESS.Bus

What kind of family relationship links the ACCESS.Bus to the I²C bus? We can assure you that the relationship is direct electronically, and that they are first protocol cousins.

We'll bring you up to date about the latest family news on ACCESS.Bus, and its relationship to I²C, in a later book.

# 3

# The I²C Protocol

## Part 1

## The I²C Bus in Standard (100 kHz) Mode

The first two chapters introduced you to microcontroller-controlled buses. It's now time to discuss in detail the bus standard known as I²C (an abbreviation for 'Inter-Integrated Circuit Bus'). As we've already demonstrated, the I²C bus standard offers many of the features you'll need to meet controls bus requirements.

This chapter describes and analyzes in depth the protocol that governs I²C bus operation. We've taken steps to prevent this discussion from becoming overly dry and academic. Although this is a serial bus story, we have presented a concrete application in 'parallel', to show you how to translate engineering theory into real-world practice.

Chapter 13 which is dedicated to development tools also shows how to generate the I²C bus and its messages with a stripped-down version of I²C, the deluxe version, and a number of variations in between.

The best way to start is to make sure we are on the same wave length. To stay in sync, we need to choose an approximate vocabulary for our discussion.

## I²C Bus Terminology

Transmitter:      the device that sends data to the bus. A transmitter can either be a device which puts data on the bus of it's own accord (a 'master-transmitter'), or in response to a request for data from another device (a 'slave-transmitter').

Receiver:              the device that receives data from the bus.

Master:                the component that initializes a transfer, generates the
                       clock signal and terminates the transfer. A master can be
                       either a transmitter or a receiver.

Slave:                 the device addressed by the master. A slave can be either
                       receiver or transmitter.

Multi-master:          the ability for more than one master to co-exist on the
                       bus at the same time without collision or data loss.

Arbitration:           the prearranged procedure that authorizes only one
                       master at a time to take control of the bus.

Sychronization:        the prearranged procedure that synchronizes the clock
                       signals provided by two or more masters.

SDA:                   data signal line (Serial DAta).

SCL:                   clock signal line (Serial CLock).

## Terminology for bus transfer

F (FREE):              the bus is free; the data line SDA and the SCL clock are
                       both in the **high** state.

S (START):             data transfer begins with a **start condition** (*not* a start
                       bit). The level of the SDA data line changes from **high** to
                       **low,** while the SCL clock line remains **high**. When this
                       occurs, the bus is '*busy*.'

C (CHANGE):            while the SCL clock line is low, the data bit to be
                       transferred can be applied to the SDA data line by a
                       transmitter. During this time, SDA may change it's state
                       as long as the SCL line remains low.

D (DATA):              a **high** or **low** bit of information on the SDA data line is
                       valid during the **high** level of the SCL clock line. This
                       level must be maintained stable during the entire time
                       that the clock remains high to avoid misinterpretation as
                       a Start or Stop condition.

P (STOP):           data transfer is terminated by a **stop condition**, (*not* a stop bit). This occurs when the level on the SDA data line passes from the low state to the high state, while the SCL clock line remains high. When the data transfer has been terminated, the bus is free once again.

# Hardware Configuration of Devices Connected to the Bus

Figure 3.1 shows examples of how the electrical circuit diagrams are configured of the input/output devices physically connected to the bus.

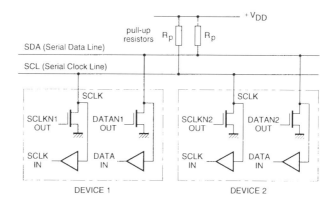

**Figure 3.1**

Many of the I²C bus' strengths result from this single choice of configuration! To see why, let's take a magnifying glass and look a little closer.

The two SDA and SCL lines are each bi-directional (both input and output) and are connected to the positive power supply line via resistors that serve a double function of load and pull-ups. At rest, when the transistors are not conducting, the bus is said to be **free**; these two lines are then released to the high state via the pull-up resistors.

The choice of this output configuration (open collector or open drain ) allows the realization of the **wired-AND** function for pins of the same name. (All other components connected to the bus must also allow the same function to be realized, i.e. open-collector or open-drain).

This output configuration allows any transmitting device to drive the bus while simultaneously monitoring the bus level to verify whether the intended information has indeed been put on the bus.

Does this make sense to you? Thanks to this fancy footwork, it is possible, to achieve synchronization and arbitration between different masters, allowing multi-master capability on the I²C bus!

# Consequences of the Configuration

## Load capacity

Load capacity, or the number of circuits that can be simultaneously connected to the bus, generates a number of issues.

### Static load capacity

To deal with *static load* capacity on the I²C bus, we have to extend our thinking. We have to be prepared for a worst case scenario – we don't know how many of the various circuits connected to the bus will be electrically operational, how they are connected, or when they are connected.

This worst case scenario corresponds to the maximum current load that the output transistor can handle, whether its technology is bipolar or CMOS. The maximum load current in the specification is 3 mA. With a power supply voltage of 5 volts, we can calculate the minimum value of the pull-up resistors (all of which are in parallel) as 1.5 kΩ. The maximum value is not critical at this time.

### Dynamic load capacity

A number of parameters define *dynamic load* capacity:

- the maximum value of the rise time (1 microsecond) is allowed for the signals on SDA and SCL. (This value is principally related to the maximum data flow rate – 100 kbits/s , but it is totally independent of the data rate actually used);

- there may be more than one master connected to the bus, operating at the maximum rate;

- the maximum allowed value of the output capacitance of an elementary circuit (20 pF).

If we only take into account the maximum value of the rise time, it is easy to calculate the maximum capacitance that can be loaded onto the bus. The curve in Figure 3.2 shows the results of these calculations.

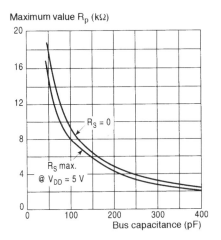

**Figure 3.2**

We can (proudly) conclude that 400 pF represents a maximum load capacity of 19 (+1 for the master) circuits of 20 pF maximum capacitance, or that we can put one circuit (of 20 pF max) at the end of a 3.6 m cable with 100 pF/m capacitance!

This kind of reasoning, however, is simplistic. Everyone knows that:

+ all of the circuits will not have the maximum 20 pF value, (7 pF is most common for CMOS ICs)

+ it is easy to reduce the rise time by buffering the outputs, thus decreasing the apparent values of the output resistances. (See Chapter 11 concerning bus extensions for more on this subject).

## High and Low Bus Levels

Before we discuss data transfer, we need precise information about high and low electrical values on the bus.

Because many types of components that use various technologies (NMOS, CMOS, bipolar ), can be connected the 0 (low) and 1 (high) logic levels of the bus have to be determined in a way that doesn't penalize any of these possibilities. For this reason, we selected the following values.

For a 5 volt power supply voltage:

**Vil max = 1.5 V and Vih min = 3.0 V**

For CMOS circuits, which accept a wide range of power supply voltage:

**Vil max = 0.3 VDD and Vih min = 0.7 VDD**

Whatever the power supply voltage:

**Vol max = 0.4 V at 3 mA sink current**

When you study these values carefully, you may notice that there is no relationship to the power supply voltages of the integrated circuits connected to the bus (5 V, 12 V, 15 V).

**Important remark:**
In the case of multiple power supply voltages, it's certainly a good idea to add diodes, connected head to tail (see Figure 3.3), to the outputs (open collectors/drains) to protect against possible over-voltages.

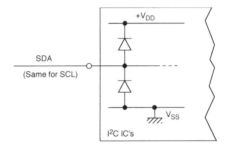

**Figure 3.3**

But, what happens if these multiple power supplies are physically separated (Figure 3.4) and one of the modules is **unpowered** (its power supply voltage is zero from a source whose internal resistance is low, as seen Figure 3.5)? One of the diodes will go into conduction, permanently forcing the bus to a low level and blocking it. For this reason I²C circuits which are often used with a seperate (backup) supply (example: battery backup) such as microcontrollers and real time clocks are often not provided with internal protection diodes to $V_{DD}$. It is up to the system designer to include other types of protection against over-voltages that

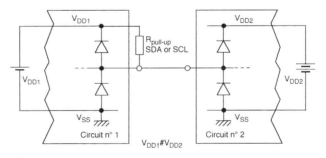

**Figure 3.4**

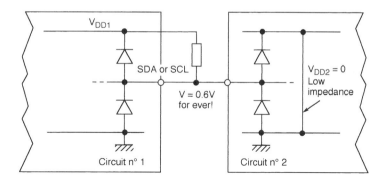

**Figure 3.5**

may appear on their power supply lines. Now, we can discuss data transfer on the bus.

# Data Transfer on the Bus

## Information transfer

Information transfer is closely related to the I²C bus protocol. Like all structured protocols, the I²C's protocol includes:

+ definitions of the high and low electrical levels

+ timing specifications

+ max data rate

+ operating conditions

+ conditions for changes of state

+ conditions for data validity

+ START and STOP conditions

+ word formats

+ transmission formats

+ acknowledgment procedures

+ synchronization procedures

+ arbitration procedures

♦ Transfer operating conditions
(Basic principle: a clock pulse is generated each time a bit is transferred).

♦ Conditions for changes of state and validity of data: these two conditions are intimately related and it is reasonable to take the SCL clock line as reference. In this case, you define (Figure 3.6):

Data validity:

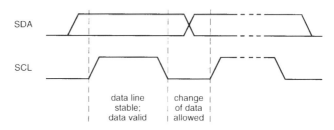

**Figure 3.6**

Data are considered to be valid when the SDA data line is **stable** (high or low) while the SCL clock line is in its high state.

Change of state:

During data transmission, regardless of the level of the SDA data line, this level is only allowed to change when the SCL clock line is in its **low** state.

## Start and Stop conditions (Figure 3.7)

Within the I²C definition protocol, the situations that determine start and stop conditions are *unique*. Take note that they are start and stop **conditions** and *not* start and stop bits.

**Start condition:**

This condition occurs *only* when the SDA data line passes from the **high** state to the **low** state while the clock line remains in the **high** state.

**Stop condition:**

This condition occurs *only* when the SDA data line passes from the **low** state to the **high** state while the clock line remains **high**.

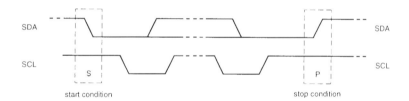

**Figure 3.7**

Note that *Start* and *Stop* conditions are the only times in the I²C bus protocol when the SDA line is allowed to change states while the SCL line is high.

**Bear in mind**: the **start** and **stop** conditions are *always* generated by a bus master. The bus is considered busy after the start condition, and the bus is considered free after a **stop** condition has been completed.

# Transmission Formats

## Word formats

Let's take a look at word format during data transfer. Each word impressed on the SDA data line must be 8 bits long. This allows the words to be directly processed by a standard 8-bit microcontroller as they are received. The data contained in the word are transferred with the most significant bit (MSB) first. We will define the meaning of each of the transmitted words when we discuss transmission formats.

## Transfer formats

### Basic principle of a transfer (Figure 3.8)

Each transfer begins with a START condition. Each word to be transmitted must be followed by an acknowledgment bit (see below). This is transmitted from slave to master during a 9th clock pulse.

Note that the I²C protocol allows a receiving device to force a transmitter to momentarily interrupt its transmission in order to accomplish a higher priority task. To indicate this situation to the transmitter, the receiver forces (and maintains) the SCL clock line to its **low** level. This immediatly arrests bus activity. To continue the message, the receiver simply releases the clock line indicating to the transmitter that it can continue its transfer.

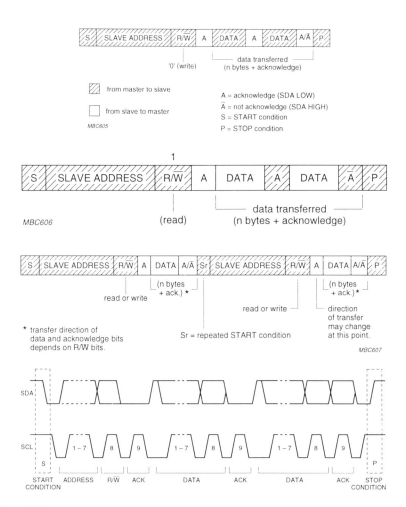

**Figure 3.8**

The number of words transmitted in the course of a single transfer is technically unlimited, but practically set by a device's function. Each transfer ends with a STOP condition, generated by the master.

## The meaning of each word during transmission

### The first word transmitted

The first word transmitted is the address of the slave that the master has selected. This address is seven (7) bits long.

The eighth bit (LSB) of the first word has a special meaning. It is called the R/$\overline{\text{W}}$ bit.

The R/$\overline{\text{W}}$ bit indicates the direction of data transfer on the bus:

  ◆ **zero** indicates a write operation from master to slave

  ◆ **one** indicates a read operation, (to read data contained in a slave)

When an address is transmitted, each of the components physically present on the bus compares the seven first bits following the start condition to its own address. If the addresses match exactly, it considers itself addressed. It then obediently waits for the eighth (R/$\overline{\text{W}}$) bit to determine if it will be a slave-receiver (one that will be written to) or a slave-transmitter (which will be asked to pour its little heart out!).

Let's take a detailed look at the configuration of the address bits. Technically, you can choose anything you want – However practically slave addresses are allocated according to certain sensible guidelines.

To avoid total address-anarchy, an I²C Committee (usually) coordinates allocations to the I²C club members. (See the example in Table 3.1).

The same type of circuit is often used more than once (EEPROM, for example) in the same system. This means that unique addresses for circuits of the same type are required so you can independently access them. To solve this quandary, one part of the seven bit address field is often fixed (usually the most significant bits 4 and 5) and the others are **modifiable**.

We haven't said anything yet about the way in which they can be modified. These bits can be modified, for example, either by hardware (digital or analog) or by software or . . . however you like! Usually the programmable address bits are brought out as pins on the IC. How many variable bits you can have then depends more on the number of available pins on the package than on anything else. For example, a circuit that has four fixed address bits and three bits to be configured gives the designer the opportunity to place an additional seven little brothers of a given circuit on the same bus.

We have predicted some questions and objections:

1. With this protocol and seven address bits you can *only* address 128 circuits whose addresses are different. Is this enough? For 99.9% of applications, yes! For the remaining 0.1%, there is no need to worry. We will provide lots of ideas to increase the address range. For example, if you want to make superb giant LED or LCD displays , you'll have plenty of opportunity to satisfy their unquenchable thirst for integrated circuits.

**Table 3.1** Philips semiconductors assigned I²C-bus addresses

| Type number | Description | I²C slave address | | | | | | |
|---|---|---|---|---|---|---|---|---|
| | | A6 | A5 | A4 | A3 | A2 | A1 | A0 |
| – | General call address | 0 | 0 | 0 | 0 | 0 | 0 | 0 |
| – | Reserved addresses | 0 | 0 | 0 | 0 | X | X | X |
| NE5751 | Audio processor RF communication | 1 | 0 | 0 | 0 | 0 | 0 | A |
| PCA1070 | Programmable speech transmission IC | 0 | 1 | 0 | 0 | 0 | 1 | A |
| PCA8510 | Stand-alone OSD circuit | 1 | 0 | 1 | 1 | 1 | 0 | 1 |
| PCA8516 | Stand-alone OSD IC | 1 | 0 | 1 | 1 | 1 | 0 | 1 |
| PCA8581/C | 128 × 8-bit EEPROM | 1 | 0 | 1 | 0 | A | A | A |
| PCB5020 | Digital audio signal processor for car radio including ROM | 0 | 0 | 1 | 1 | 0 | A | A |
| PCB5021 | Digital audio signal processor for car radio excluding ROM | 0 | 0 | 1 | 1 | 0 | A | A |
| PCD3311C | DTMF/modern/musical-tone generator | 0 | 1 | 0 | 0 | 1 | 0 | A |
| PCD3312C | DTMF/modern/musical-tone generator | 0 | 1 | 0 | 0 | 1 | 0 | A |
| PCD4430 | Programmable tone detector and DTMF generator | 0 | 1 | 0 | 0 | 0 | 0 | A |
| PCD4440 | Analog voice scrambler/descrambler for mobile telephones | 1 | 1 | 0 | 1 | 1 | 1 | A |
| PCF5002 | Pager decoder | 0 | 1 | 0 | 0 | 1 | 1 | 1 |
| PCF1810 | 8 × 8 cross-popint matrix analog switch | 0 | 0 | 1 | 1 | 1 | A | A |
| PCF2116 | LCD controller/driver | 0 | 1 | 1 | 1 | 0 | 1 | A |
| PCF8566 | 96-segment LCD driver 1:1–1:4 Mux rates | 0 | 1 | 1 | 1 | 1 | 1 | A |
| PCF8568 | LCD row driver for dot matrix displays | 0 | 1 | 1 | 1 | 1 | 0 | A |
| PCF8569 | LCD column driver for dot matrix displays | 0 | 1 | 1 | 1 | 1 | 0 | A |
| PCF8570/C | 256 × 8-bit static RAM | 1 | 0 | 1 | 0 | A | A | A |
| PCF8573 | Clock/calendar | 1 | 1 | 0 | 1 | 0 | A | A |
| PCF8574 | 8-bit remote I/O port (I²C-bus to parallel converter) | 0 | 1 | 0 | 0 | A | A | A |
| PCF857A | 8-bit remote I/O port (I²C-bus to parallel converter) | 0 | 1 | 0 | 0 | A | A | A |
| PC8576 | 16-segment LCD driver 1:1–1:4 Mux rates | 0 | 1 | 1 | 1 | 0 | 0 | A |
| PCF8577A | 32/64-segment LCD display driver | 0 | 1 | 1 | 1 | 0 | 1 | 1 |
| PCF8577C | 32/64-segment LCD display driver | 0 | 1 | 1 | 1 | 0 | 1 | 0 |
| PCF8578 | Row/column LCD dot matrix driver/display | 0 | 1 | 1 | 1 | 1 | 0 | A |
| PCF8579 | Row/column LCD dot matrix driver/display | 0 | 1 | 1 | 1 | 1 | 0 | A |
| PC F8582/A | 256 × 8-bit CMOS EEPROM | 1 | 0 | 1 | 0 | A | A | A |
| PCF8583 | 256 × 8-bit RAM/clock/calendar | 1 | 0 | 1 | 0 | 0 | 0 | A |
| PCF8591 | 4-channel, 8-bit Mux ADC and one DAC | 1 | 0 | 0 | 1 | A | A | A |
| PCF8593 | Lower-power clock calendar | 1 | 0 | 1 | 0 | 0 | 0 | 1 |
| PCX8594X-2 | 512 × 8-bit CMOS EEPROM | 1 | 0 | 1 | 0 | A | A | P |
| PCX8598X-2 | 1024 × CMOS EEPROM | 1 | 0 | 1 | 0 | A | P | P |
| SAA 1064 | 4-digit LED driver | 0 | 1 | 1 | 1 | 0 | A | A |
| SAA1136 | PCM audio interface | 0 | 0 | 1 | 1 | 1 | 0 | 0 |
| SAA1137 | PCM audio prtocessor | 0 | 1 | 0 | 0 | 0 | 0 | A |
| SAA1300 | Tuner switch circuit | 0 | 1 | 0 | 0 | 0 | A | A |
| SAA1770 | D2MAC decoder for satellite and cable TV | 0 | 0 | 1 | 1 | 1 | 1 | A |
| SAA2502 | MPEG audio source decoder | 0 | 0 | 1 | 1 | 1 | 0 | 1 |
| SAA2510 | Video-CD MPEG-audio/video decoder | 0 | 0 | 1 | 1 | 0 | 1 | A |

**Table 3.1**  *(continued)*

| Type number | Description | A6 | A5 | A4 | A3 | A2 | A1 | A0 |
|---|---|---|---|---|---|---|---|---|
| | | | | $I^2C$ slave address | | | | |
| SAA4700 | VPS dataline processor | 0 | 0 | 1 | 0 | 0 | 0 | A |
| SAA5240 | 625-line teletext decoder; english/german/swedish | 0 | 0 | 1 | 0 | 0 | 0 | 1 |
| SAA5240B | 625-line teletext decoder; french/italian/german | 0 | 0 | 1 | 0 | 0 | 0 | 1 |
| SAA5240P/D | 625-line teletext decoder; spanish | 0 | 0 | 1 | 0 | 0 | 0 | 1 |
| SAA5240P/C | 625-line teletext decoder; arabic | 0 | 0 | 1 | 0 | 0 | 1 | 0 |
| SAA5240P/F | 625-line teletext decoder; hebrew | 0 | 0 | 1 | 0 | 0 | 0 | 1 |
| SAA5241A | 625-line teletext decoder; english/german/swedish | 0 | 0 | 1 | 0 | 0 | 0 | 1 |
| SAA5241B | 625-line teletext decoder; french/italian/german | 0 | 0 | 1 | 0 | 0 | 0 | 1 |
| SAA5243P/K | Computer controlled teletext circuit | 0 | 0 | 1 | 0 | 0 | 0 | 1 |
| SSAA5243P/L | Computer controlled teletext circuit | 0 | 0 | 1 | 0 | 0 | 0 | 1 |
| SAA5243P/H | Computer controlled teletext circuit | 0 | 0 | 1 | 0 | 0 | 0 | 1 |
| SAA5243P/E | Computer controlled teletext circuit | 0 | 0 | 1 | 0 | 0 | 0 | 1 |
| SAA5244 | Integrated VIP and teletext | 0 | 0 | 1 | 0 | 0 | 0 | 1 |
| SAA5245 | 525-line teletext decoder/controller | 0 | 0 | 1 | 0 | 0 | 0 | 1 |
| SAA5246 | Integrated VIP and teletext | 0 | 0 | 1 | 0 | 0 | 0 | 1 |
| SAA5252 | Line 21 decoder | 0 | 0 | 1 | 0 | 1 | 0 | 0 |
| SAA7110 | Digital multistandard decoder | 1 | 0 | 0 | 1 | 1 | 1 | A |
| SAA7140 | High performance video scaler | 0 | 1 | 1 | 1 | 0 | 0 | A |
| SAA7151B | 8-bit digital multistandard TV decoder | 1 | 0 | 0 | 0 | 1 | A | 1 |
| SAA7152 | Digital comb filter | 1 | 0 | 1 | 1 | 0 | 0 | 1 |
| SAA7165 | Video enhancement D/A processor | 1 | 0 | 1 | 1 | 1 | 1 | 1 |
| SAA7186 | Digital video scaler | 1 | 0 | 1 | 1 | 1 | A | 0 |
| SAA7191 | Digital multistandard TV decoder | 1 | 0 | 0 | 0 | 1 | A | 1 |
| SAA192A | Digital colour space-converter | 1 | 1 | 1 | 0 | 0 | 0 | A |
| SAA7194 | Digital video decoder/scaler | 0 | 1 | 0 | 0 | 0 | 0 | A |
| SAA7199B | Digital multistandard encoder | 1 | 0 | 1 | 1 | 0 | 0 | A |
| SAA7250 | General purpose digital audio signal processor | 0 | 0 | 1 | 1 | 0 | 0 | A |
| SAA7370 | CD-decoder plus digital servo processor | 0 | 0 | 1 | 1 | 0 | 0 | A |
| SAA9020 | Field memory controller | 0 | 0 | 1 | 0 | 1 | A | A |
| SAA9041 | Digital video text – backend | 0 | 0 | 1 | 0 | 0 | 0 | 1 |
| SAA9051 | Digital multistandard colour TV decoder | 1 | 0 | 0 | 0 | 1 | A | 1 |
| SAA9053 | Digital NTSC TV decoder | 1 | 0 | 0 | 0 | 1 | A | 1 |
| SAA9056 | Digital SECAM colour decoder | 1 | 0 | 0 | 0 | 1 | A | 1 |
| SAA9060 | Black and white PIP | 1 | 0 | 0 | 0 | 1 | 1 | 0 |
| SAA9065 | Video enhancement and D/A processor | 1 | 0 | 1 | 1 | 1 | 1 | 1 |
| SAB3028 | Remote control RC-5 transcoder | 0 | 1 | 0 | 0 | 1 | 1 | 0 |
| SAB3035 | Digital tuning circuit for computer-controlled TV | 1 | 1 | 0 | 0 | 0 | A | A |
| SAB3036 | Digital tuning circuit for computer-controlled TV | 1 | 1 | 0 | 0 | 0 | A | A |
| SAB3037 | Digital tuning circuit for computer-controlled TV | 1 | 1 | 0 | 0 | 0 | A | A |
| SAB9070 | PIP8 controller | 0 | 0 | 1 | 0 | 0 | 1 | 0 |
| SAF1134P | Dataline 16 decoder for VPS (gate array) | 0 | 0 | 1 | 0 | 0 | A | A |
| SAF1135P | Dataline 16 decoder for VPS (cell array) | 0 | 0 | 1 | 0 | 0 | A | A |
| TDA1551B | 2 × 22 W BTL audio power amplifier | 1 | 1 | 0 | 1 | 1 | 0 | 1 |
| TDA1551Q | 2 × 22 W BTL audio power amplifier | 1 | 1 | 0 | 1 | 1 | 0 | 0 |

*(continues overleaf)*

**Table 3.1** (continued)

| Type number | Description | I²C slave address | | | | | | |
|---|---|---|---|---|---|---|---|---|
| | | A6 | A5 | A4 | A3 | A2 | A1 | A0 |
| TDA4670 | Picture signal improvement (PSI) circuit | 1 | 0 | 0 | 0 | 1 | 0 | 0 |
| TDA4670 | Picture signal improvement (PSI) circuit | 1 | 0 | 0 | 0 | 1 | 0 | 0 |
| TDA4671 | Picture signal improvement (PSI) circuit | 1 | 0 | 0 | 0 | 1 | 0 | 0 |
| TDA4672 | Picture signal improvement (PSI) circuit | 1 | 0 | 0 | 0 | 1 | 0 | 0 |
| TDA4680 | Video processor | 1 | 0 | 0 | 0 | 1 | 0 | 0 |
| TDA4685 | Video processor | 1 | 0 | 0 | 0 | 1 | 0 | 0 |
| TDA4687 | Video processor | 1 | 0 | 0 | 0 | 1 | 0 | 0 |
| TDA4688 | Video processor | 1 | 0 | 0 | 0 | 1 | 0 | 0 |
| TDA4780 | Video control with gamma control | 1 | 0 | 0 | 0 | 1 | 0 | 0 |
| TDA6360 | Five-band equalizer for car radio | 1 | 0 | 0 | 0 | 0 | 1 | A |
| TDA8045 | QAM-64 demodulator | 0 | 0 | 0 | 1 | 1 | A | A |
| TDA8363 | Single chip NTSC decoder | 1 | 0 | 0 | 0 | 1 | 0 | 0 |
| TDA8366 | One-chip multistandard video | 1 | 0 | 0 | 0 | 1 | 0 | 1 |
| TDA8370 | High/medium perf. sync. processor | 1 | 0 | 0 | 0 | 1 | 1 | 0 |
| TDA8376 | One-chip multistandard video | 1 | 0 | 0 | 0 | 1 | 0 | 1 |
| TDA8405 | Stereo/dual language decoder | 1 | 0 | 0 | 0 | 0 | 1 | 0 |
| TDA8415 | TV/VCR stereo/dual sound processor | 1 | 0 | 0 | 0 | 0 | 1 | 0 |
| TDA8416 | TV/VCR stereo/dual sound processor | 1 | 0 | 1 | 1 | 0 | 1 | 0 |
| TDA8417 | TV/VCR stereo/dual sound processor | 1 | 0 | 0 | 0 | 0 | 1 | 0 |
| TDA8420 | Audio processor with loudspeaker and headphone channel | 1 | 0 | 0 | 0 | 0 | 0 | A |
| TDA8421 | Audio processor with loudspeaker and headphone channel | 1 | 0 | 0 | 0 | 0 | 0 | A |
| TDA8424 | Audio processor with loudspeaker channel | 1 | 0 | 0 | 0 | 0 | 0 | 1 |
| TDA8425 | Audio processor with loudspeaker channel | 1 | 0 | 0 | 0 | 0 | 0 | 1 |
| TDA8426 | Hi-fi stereo audio processor | 1 | 0 | 0 | 0 | 0 | 0 | 1 |
| TDA8432 | Sync. controller and deflection processor | 1 | 0 | 0 | 0 | 1 | 1 | A |
| TDA8433 | TV deflection processor | 1 | 0 | 0 | 0 | 1 | 1 | A |
| TDA8440 | Video/audio switch | 1 | 0 | 0 | 1 | A | A | A |
| TDA8442 | Interface for colour decoder | 1 | 0 | 0 | 0 | 1 | 0 | 0 |
| TDA8443/A | YUV/RGB matrix switch | 1 | 1 | 0 | 1 | A | A | A |
| TDA8444 | Octal 6-bit DAC | 0 | 1 | 0 | 0 | A | A | A |
| TDA8461 | PAL/NTSC colour decoder with RGB processor | 1 | 0 | 0 | 0 | 1 | 0 | A |
| TDA8466 | PAL/NTSC colour decoder with RGB processor | 1 | 0 | 0 | 0 | 1 | 0 | A |
| TDA8480 | RGB gamma-correction processor | 1 | 0 | 0 | 0 | 0 | 1 | A |
| TDA8450 | 4 × 4 video switch matrix | 1 | 0 | 0 | 1 | A | A | A |
| TDA9140 | Alignment-free multistandard decoder | 1 | 0 | 0 | 0 | 1 | A | 1 |
| TDA9141 | Alignment-free multistandard decoder | 1 | 0 | 0 | 0 | 1 | A | 1 |
| TDA9145 | Multistandard decoder | 1 | 0 | 0 | 0 | 1 | 0 | 1 |
| TDA9150 | Deflection processor | 1 | 0 | 0 | 0 | 1 | 1 | 0 |
| TDA9160 | Multistandard decoder/sync. processor | 1 | 0 | 0 | 0 | 1 | A | 1 |
| TDA9161 | Bus-controlled decoder/sync. processor | 1 | 0 | 0 | 0 | 1 | 0 | 1 |
| TDA9162 | Multistandard decoder/sync. processor | 1 | 0 | 0 | 0 | 1 | A | 1 |
| TDA9840/T | TV stereo/dual sound processor | 1 | 0 | 0 | 0 | 0 | 1 | 0 |

**Table 3.1** *(continued)*

| Type number | Description | I²C slave address | | | | | | |
|---|---|---|---|---|---|---|---|---|
| | | A6 | A5 | A4 | A3 | A2 | A1 | A0 |
| TDA9860 | Hi-fi audio processor | 1 | 0 | 0 | 0 | 0 | 0 | A |
| TEA6000 | FM/IF and search tuning interface | 1 | 1 | 0 | 0 | 0 | 0 | 1 |
| TEA6100 | FM/IF for computer-controlled radio | 1 | 1 | 0 | 0 | 0 | 0 | 1 |
| TEA6300 | Sound fader control and preamplifier/source selector | 1 | 0 | 0 | 0 | 0 | 0 | 0 |
| TEA6320 | 4-input tone/volume controller with fader control | 1 | 0 | 0 | 0 | 0 | 0 | 0 |
| TEA6330 | Tone/volume controller | 1 | 0 | 0 | 0 | 0 | 0 | 0 |
| TEA6360 | 5-band equalizer | 1 | 0 | 0 | 0 | 1 | 1 | A |
| TSA5510 | 1.2 GHz PLL frequency synthesizer without AFC ADC | 1 | 1 | 0 | 0 | 0 | A | A |
| TSA5511 | 1.3 GHz PLL frequency synthesizer for TV | 1 | 1 | 0 | 0 | 0 | A | A |
| TSA5512 | 1.3 GHz PLL frequency synthesizer for TV | 1 | 1 | 0 | 0 | 0 | A | A |
| TSA5514 | 1.3 GHz PLL frequency synthesizer for TV | 1 | 1 | 0 | 0 | 0 | A | A |
| TSA5519 | 1.3 GHz PLL frequency synthesizer ADC | 1 | 1 | 0 | 0 | 0 | A | A |
| TSA6057 | Radio tuning PLL frequency synthesizer | 1 | 1 | 0 | 0 | 0 | 1 | A |
| TSA6060 | Radio tuning PLL frequency synthesizer | 1 | 1 | 0 | 0 | 0 | 1 | A |
| TSA6061 | 150 MHz PLL and IF-counter | 1 | 1 | 0 | 0 | 0 | 1 | A |
| UMA1000T | Data processor for mobile telephones | 1 | 1 | 0 | 1 | 1 | A | A |
| UMA1009 | Frequency synthesizer for mobile telephones | 1 | 1 | 0 | 0 | 0 | A | A |
| UMA1010 | Frequency synthesizer for mobile telephones | 1 | 1 | 0 | 0 | 0 | A | A |
| UMA1014T | Frequency synthesizer for mobile telephones | 1 | 1 | 0 | 0 | 0 | 1 | A |
| – | Reserved addresses | 1 | 1 | 1 | 1 | X | X | X |

X = Don't care
A = Programmable address bit
P = Page selection bit

2. With three programmable bits, you cannot have more than eight circuits of the same type on the bus. However, there are a number of tricks for connecting many more by time multiplexing these **modifiable** bits. Certain simple configurations allow more than 300 circuits of the same type to be addressed! (See the chapter on I²C input/output modules). Just for laughs, you can use the contents of the message transmitted to a circuit to tell it to reconfigure its own address! (We have even succeeded in driving circuits totally mad so that they no longer know their own names!)

## Reserved addresses

In the first transmitted byte there are also reserved addresses that correspond to particular multi-master protocol functions , such as 'general call' address, and the 'start word' (which we will examine at the end of this chapter).

## The following words

The significance of the words that follow the address word are device-dependent. They are coded by 8 (eight) bits per word. Their function is to carry sub-addresses, data, control bytes, etc. as defined by each device's specification. Each 8-bit word is punctuated by an acknowledge bit.

The number of words transmitted is not limited (in principle). It depends on the type of circuit addressed, and on the reality that you have to stop somewhere. If you keep on sending data to one receiver, you'll never do anything else!

# Acknowledgment

As you can see in Figure 3.9, an acknowledgment procedure is introduced after every word transmitted. To properly understand this function and its power, it is necessary to thoroughly analyse its principle.

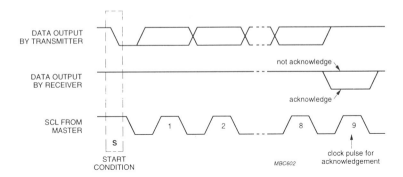

**Figure 3.9**

## Acknowledgement conditions

•  Data transfer with acknowledgment is **compulsory**

•  an acknowledgment must take place after each word is transmitted

•  the ninth clock pulse (during which the receiving component must send its acknowledgment back to the transmitter) is generated by the master in control of the bus at the time

•  during the ninth clock pulse the transmitter must leave its SDA data line free after **releasing** it to the **high** state.

For a positive **acknowledgment** to be valid, the receiver must pull the SDA data line to the **low** state while the SCL clock line is **high** and stable.

## Non-acknowledgment

1. We need to have well-defined action procedures for all cases of non-acknowledgment. Here are two specific examples: The receiver is involved in the completion of a real time task (writing to an E2PROM, for example) and can't accept the data offered. To be nasty to the transmitter, the receiver purposely leaves the SDA data line in the state in which it found it, (the **high** state). The master understands this as a refusal (non-acknowledgment). It has all the time it needs to generate a **stop** condition to abandon this recalcitrant receiver and take care of another, more accommodating receiver. Since these regrettable incidents usually don't last long, the master can periodically query the receiver's state of mind to continue or restart the transfer. In reality, this is a common scenario with EEPROMs; during programming, an EEPROM is not able to take in any additional data until it has finished it's internal write cycle.

2. A second case is common but a little more complicated. The master that generates the clock is the receiver. For example, it has just asked a memory to send its contents. Because of its other responsibilities, such as managing an interrupt, the master may want the slave transmitter to understand that it wants the transfer stopped. By not acknowledging the last word transmitted by the slave, the master tells the slave to release the receiver SDA data line so it can generate a **stop** condition.

## A short time-out

To conclude our introduction of basic principles, we're going to take a brief time-out. The following example demonstrates what happens in an actual circuit.

As the basis of this little interlude, we have chosen the LED display circuit SAA 1064, which controls a four digit (seven segment plus decimal point) LED display. It is easy to use and its function is mainly **passive**. SAA 1064 acts as a **slave-receiver** in ninety-nine percent of its applications; its only function is to write. You may be wondering what the other one percent is good for.

A display circuit can have a tiny bit of intelligence when it has been designed to let you know that someone has cut off its power supply. To achieve this, the LED circuit becomes a slave-transmitter and confesses that it has nothing to display (during power interruption the IC's lose their information contents).

An SAA 1064 'family' (with 2 programmable address bits there may be quite a few in a large display!) goes by the charming name of: **01110 XX**(.).

As you may have noticed, we divided its address into three parts:

+ 01110 internally fixed part for the generic family SAA 1064

+ XX configurable part of the circuit which, for example, we will choose equal to: 10

+ (.) this is the R/$\overline{\text{W}}$ bit (R = 1; W = 0). In our case (write), this bit will be 0.

Therefore:       specific address       0111010.
                      write operation                0
Full byte to be transmitted    0111 0100    or 76 (in hexadecimal)

The specification of the circuit SAA 1064 (Figure 3.10) shows you that the following bytes represent an instruction word first, then a control word. The instruction word corresponds for this circuit to a sub-address whose function we will discuss later. For the sake of simplicity, let's assume that this word is equal to 0 0 in hexadecimal.

The control word describes the visual aspect that the user specifies. In the case of a dynamic duplexed display, at maximum current circulating in the display, this word should be: **X111 0111** or 77 in hexadecimal.

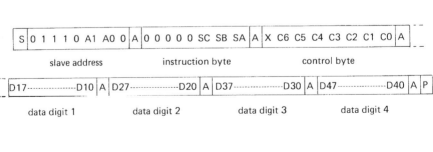

Figure 3.10

The following four words represent the values that each digit should display.

The contents of the transmission sequence are as follows:

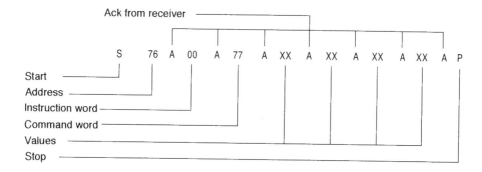

This long preamble should help you to move towards designing and implementing I²C bus-controlled systems. Now that we've forced you to assimilate all of these basic principles of how and why this serial bus systems works so well, you are ready to immerse yourself completely into I²C. After this warm-up, you're ready to shift gears and move up to higher data rates!

# Part 2
# The I$^2$C Bus in High-Speed Mode

The original I$^2$C bus protocol we have just presented is still in force and will be for quite some time. But new things are happening in the I$^2$C kingdom – a technological revolution.

For several years, rumors of improvements have come and gone. Many new so-called 'serial buses' have appeared on the market, a host of Johnny-come-lately's hoping to dethrone the reigning standard.

Late-comers to the scene have the advantage of being able to criticize and improve, but still face enormous handicaps. As we have already remarked there are more than 300 circuits (with different, complementary or competitive functions, depending on the manufacturer) that are compatible with the I$^2$C bus.

This represents a considerable advantage, but also makes it difficult to introduce evolutionary changes in the protocol. The design challenge is to implement an evolution of the standard, not a revolution of the established protocol, to assure upward and downward software compatibility.

Other equally important parameters have emerged, such as EMI problems. If we need to make changes, why not kill at least three or four birds with one stone? That is our next assignment.

## New Complementary Specifications for the I$^2$C Bus

Philips Inc. published official new specifications for the I$^2$C bus in January 1992. 1995 update is available.

## Bus data rate: introducing FAST I$^2$C

Bus data rate, a more precise term than speed, has been advanced from 100 to 400 kbits per second – a four hundred percent improvement.

Some of the newer buses already operate at 1 Mbits/s, but can be difficult to manage. Some I$^2$C microcontrollers such as the 80 C552, C652 (to be discussed later), used to operate with 12 or 16 MHz clocks and delivered 250 kbits/s to the I$^2$C bus. They are now capable of running at 20, 24 and soon 30 MHz, and can use the I²C bus at 400 kbits/s and more.

One minor detail – the faster you go, the more noise you radiate!

## Data rate and EMI

The I²C bus was forced to muzzle its rise time to avoid perturbing neighbouring components, slowing it down as needed to reduce the electromagnetic radiation generated. (Experiment – turn on any bus you choose in the neighborhood of an RF receiver with the usual ground connections of doubtful quality. You will understand the problem instantly).

To avoid this, the (SDA)output stages and (SCL) clocks have been equipped with devices to control the rising and falling signal slopes. Figure 3.11 shows different examples of devices designed for integrated circuits (CMOS or bipolar).

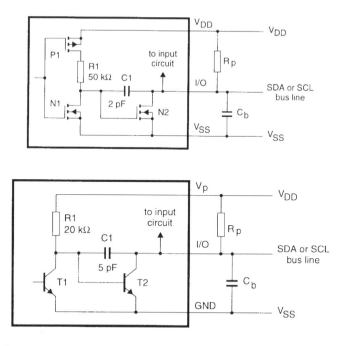

**Figure 3.11**

The slopes of the falling edges are mainly defined by the integrating network made up of Miller effect C1 capacitances and R1 resistances. Notice that the fall time is only slightly influenced by the load

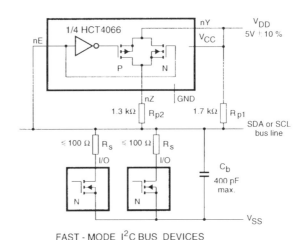

FAST - MODE I²C BUS DEVICES

**Figure 3.12**

capacitance of the Cb bus and Rp pull-up resistances. On the other hand, these two elements have a dominating influence on the rise time of the bus signals.

Nevertheless, the bus still has to meet acceptable speed requirements and we can't accept signals that take three days to rise!

An accelerating switched pull-up network (Figure 3.12) allows the reduction of the apparent dynamic value of the pull-up resistances at the moment of the low-to-high transition, decreasing the value of the RpCb time constant and providing a faster signal rise even in the presence of greater bus capacitance. This design solution controls the rise time and noise pollution.

# Determination of External Components

## Rp

You can determine minimum and maximum values of the Rp and Rs resistances for the use of the I²C bus in fast mode (400 kbits/s) with the help of Figure 3.13.

Since the fast mode I²C bus has a lower maximum rise time than before (300 ns rather than 1000 ns), at the same bus capacitance, the value of Rp has to be reduced. (Figure 3.14 makes the comparison).

The I²C standard specification requires that the output current of 3 mA must not be exceeded. Taking into account the low state , a voltage drop within the circuit of 0.4 V obtains a minimum value for Rp of 1.7 kΩ, resulting in a maximum bus capacitance of 200 pF in fast mode (300 ns).

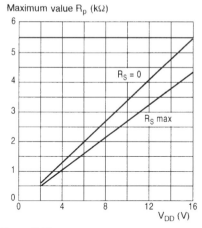
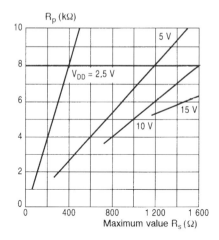

**Figure 3.13**

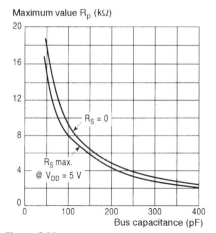
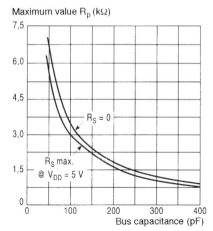

**Figure 3.14**

To solve this problem, the new proposed components specified for the I²C fast mode have a maximum current of 6 mA and the old circuits (3 mA max.) can have external switched pull-ups added for shorter rise time.

An external switched pull-up configuration is shown in Figure 3.12 as a concrete example, using an HCT 4066 that places another resistance of 1.3kΩ in parallel with Rp1 (1.7 kΩ, naturally) when the bus changes state.

# Rs

This little resistance is frequently the source of teething problems at the input gates of non-protected microcontrollers. Considered optional, this resistance's function is to protect the I²C input/output stages against parasites that circulate on the lines. These lines are, by their nature, magnificent inductive loops just waiting for the slightest little di/dt to remind you of the basic rules of electromagnetism with their EMI's induced by L(di/dt).

Rs is also there to minimize intermodulations between lines and undershoots of signals present on the bus. Its maximum value is only limited by the maximum voltage drop permitted at its terminals when the bus is switched to the low state by another component on the bus. Generally, its value doesn't exceed 100 Ω.

## The new address field

Until recently, the I²C address field (without tricks or 'dangerous' acrobatic manoeuvres) was limited structurally to 128 (7 address bits). From the very beginning, some of these addresses were reserved for future uses . If you follow the rules scrupulously, you can only use 112 truly available addresses (see Tables 3.2 and 3.3).

**Table 3.2**    Definition of bits in the first byte

| Slave address | R/W bit | Description |
|---|---|---|
| 0000 000 | 0 | General call address |
| 0000 000 | 1 | START byte |
| 0000 001 | X | CBUS address |
| 0000 010 | X | Address reserved for different bus format |
| 0000 011 | X | |
| 0000 1XX | X | Reserved for future purposes |
| 1111 1XX | X | |
| 1111 0XX | X | 10-bit slave addressing |

With the family growing every day, it was high time to define a protocol modification to take care of the new arrivals. For this reason, the address field for FAST I²C was increased from 7 to 10 bits (1024). The I²C bus Committee continues to assign addresses to license holders (see Chapter 4). Microcontroller users who are out of the fold will continue to name their components as they see fit, at the risk and peril of treading on the toes of other assigned addresses.

One question remains: where and how to introduce these ten bits into the existing I²C bus bit stream? The suspense is practically unbearable...

**Table 3.3**  I²C Address allocation table

| A6–A3 | \ A2–A0 \ 0 | 1 | 2 | 3 | 4 | 5 | 6 | 7 |
|---|---|---|---|---|---|---|---|---|
| 0 | General call address | Reserved— | ——— | ——— | ——— | ——— | ——— | → |
| 1 | | | | | | | | |
| 2 | PCF8200* | SAA5243* | | | SAA9020* | ——— | ——— | → |
|  | SAF1135:— | ——— | ——— | → | | | | |
|  | | SAA5245* | SAA9068!— | → | | | | |
| 3 | | | | | | | SAA1136! | |
| 4 | SAA1300!— | ——— | ——— | → | | | | |
|  | TDA844!— | ——— | ——— | ——— | ——— | ——— | | → |
|  | PCF8574!— | ——— | ——— | ——— | | | ——— | → |
|  | | | | | PCD3311/A!— → | | SA3028 | |
|  | | | | | PCD3312!— → | | | |
| 5 | | | | | | | | |
| 6 | | | | | | | | |
| 7 | | | PCF8577! | PCF8577A! | PCF8578* → | | PCF8566 | → |
|  | | | | | PCF8579* → | | !— | |
|  | PCF8576!— | → | | | | | | |
|  | PCF8574*— | ——— | ——— | ——— | ——— | ——— | | → |
|  | SAA1064!— | ——— | ——— | → | | | | |
| 8 | TDA8420!— | → | TDA8045* | | TDA8422! | SAA9050* | SAA9062* | |
|  | TDA8421!— | → | | | TDA8461!— | → | SAA9063* | |
|  | TEA6300/T!— | TDA8425! | | | | SAA9051* | SAA9064* | |
|  | TEA6310T!— | | | | | | | |
| 9 | TDA8440!— | ——— | ——— | ——— | ——— | ——— | ——— | → |
|  | PCVF8591* | ——— | ——— | ——— | ——— | ——— | ——— | → |
| A | PCF8583*— | → | | | | | | |
|  | PCF8570*— | ——— | ——— | ——— | ——— | ——— | ——— | → |
|  | PCF8571*— | ——— | ——— | ——— | ——— | ——— | ——— | → |
|  | PCF8572*— | ——— | ——— | ——— | ——— | ——— | ——— | → |
|  | PCF8582A!— | ——— | ——— | ——— | ——— | ——— | ——— | → |
| B | PCF8570* | ——— | ——— | ——— | ——— | ——— | ——— | → |
| C | TSA5511*— | ——— | ——— | → | | | | |
|  | SAB3035*— | ——— | ——— | → | | | | |
|  | SAB3036*— | ——— | ——— | → | | | | |
|  | SAB3037*— | ——— | ——— | → | | | | |
|  | | TEA6000*— | TSA6057!— | → | | | | |
|  | | TEA6100*— | | | | | | |
|  | TDA8400!— | ——— | ——— | → | | | | |
| D | TD8433!— | ——— | ——— | ——— | ——— | ——— | ——— | → |
|  | TDA8443A!— | ——— | ——— | ——— | ——— | ——— | ——— | → |
|  | TDA8573* | ——— | ——— | → | | | | |
| E | | | | | | | | |
| F | Reserved— | ——— | ——— | ——— | ——— | ——— | ——— | → |

\* = R/W, ! = W, : = R

# The new I²C bit stream

Figure 3.15 provides the answers.

The structure of the bit stream has been significantly modified, adding a second address byte. If you scrupulously respected the preceding rules,

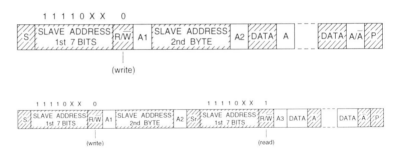

**Figure 3.15**

you would have noticed that after the START condition, the contents of the first byte transmitted (known as the address byte) contained certain values reserved for future applications; notably, values of the '1111 Xxx.' type.

The '1111 0xx.' value has been requisitioned to serve as a byte announcing that the address will be sent in 10 bits. The two 'xx' remaining are the two most significant bits of the 10 bit word. It is probable that a certain number of circuits (only those with 10 bit addresses) will respond at the first acknowledge (As1). Its companion, the value '1111 1xx.', remains in limbo waiting for future improvements.

The last bit of the first byte (the eighth bit corresponding to the '.') indicates the direction of the exchange (read or write), which leaves room to send a complete second byte with the least significant bits of this 10 bit address. At this stage, only the chosen one will recognize his address and have the right to sound his acknowledge (A2). Then, just like a parade, all the little data bits can come pouring out. Figure 3.16 shows examples of situations that may occur during these exchanges.

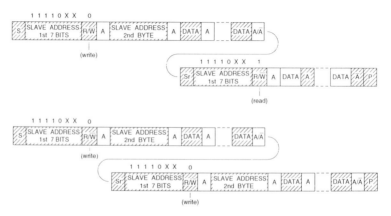

**Figure 3.16**

## Compatibility with the old mode

Is it possible to have hidden upward and downward compatibility's? Fortunately, it is possible, given the number of 'old' circuits still in use!

You have discovered that the condition presiding over the transmission of an address coded in 10 bits requires that the first (address) byte should begin with '1111 0...'. Keep in mind that all addresses that don't begin with F hexadecimal are coded in 7 bits. There is no need to panic at F hexadecimal! You need only to execute a short test to determine if a 7 or 10 bit address is coming your way. This requires a few lines of code, assuring compatibility at little cost.

The final figures show cases that can occur where a communication system includes a combination of two I²C circuits (or products), one with addresses coded in 7 bits and the other in 10 bits. Sometimes, you need to be a juggler as well as an engineer, as illustrated by Figure 3.17.

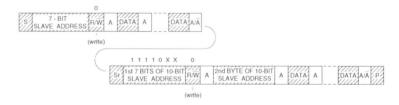

**Figure 3.17**

You have to do some fancy footwork to read a slave whose address is coded in 10 bits – first send out a 7-bit address with a write bit, then the remaining 7 bits of address plus a read bit to start the operation. (Watch out when you're designing software with downward compatibility!)

N.B. Points to remember:

1. All I²C integrated circuits existing today (7 address bits, 100 kbits/s) will *not necessarily* be systematically promoted to fast mode.
2. Any I²C circuit can technically be designed or redesigned for high speed and 10 address bits, if its function (and market potential) requires it.

## Let's wire!

When you're designing a high speed bus, you should always consider issues such as parasitic capacitance, interference between signals, crosstalk, pick-up pulses, etc. The I²C bus is sensitive in the high state because of the high impedance condition. (Usually, integrated circuits in this condition should be in the open collector/drain to ensure a wired **AND** capable of managing bus conflicts). We can state the problem somewhat differently. Let's look at some of the situations that may occur.

## Short distance on a given card

The power line ($V_{DD}$) is present just about everywhere on your card to power the various integrated circuits, and everything's fine. The problems that remain are crosstalk between the SDA and SCL lines, plus differences in their signal rise times. In these circumstances, Figure 3.18(A) shows that it is preferable to place the ground return between the SDA and SCL lines. This equalizes the parasitic capacitances on the two lines, helps to equalize rise times, and separates them (shields them) from each other, reducing any possible crosstalk that could add unwanted transitions.

(a)                                                    (b)

**Figure 3.18**

N.B.: When dealing with multi-layer boards, it's not worth the trouble to place a ground between the SDA and SCL lines because of a ground plane (and a $V_{DD}$ plane) in the middle of the sandwich. However, separating the lines will continue to reduce crosstalk.

### Separation of circuits and/or functions via the I²C bus:

If you want to go from an I²C device n1 to a second I²C device n2 via the bus, and still maintain interdependence, transport the $V_{DD}$ power supply in addition to the ground return. If your power supplies are clean (low impedance at all frequencies!), Figure 3.18(B) illustrates the best configuration. Without this configuration, motor start-up and other strange beasts can generate nasty, short spikes. These are the bane of electronics specialists, because they can sneakily superimpose themselves on the $V_{DD}$ line and then thread their way capacitively on the SDA and/or SCL lines!

We can hear you pleading for the miracle solution to short spikes.

## Solutions to avoid spikes

Here are some techniques that have worked for us from our repertoire of engineering recipes.

### The twisted pair

Take an $I^2C$ bus (at about 400 kbits/s), add a few metres of twisted pair cable, combine the SDA and SCL signals separately with a strand of $V_{SS}$. Don't mix or marinate them together – the SDA can pick up an aftertaste of SCL, or vice versa.

### The low-budget twisted pair solution

To economize on twisted pairs, we suggest the following variation, which consists of one pair:

• SDA – ground return $V_{SS}$

and another

• SCL – power supply $V_{DD}$ (provided that the $V_{DD}$ is very clean).

Sprinkle a touch of decoupling capacitance at each end of the $V_{DD}$ and $V_{SS}$ lines.

For connoisseurs, we suggest a dab of ground but, we strongly recommended you avoid coating (shielding) the wires of the pairs before serving.

Now, let's examine shielded buses.

### Shielded buses

Connect the shield to $V_{SS}$ to minimize radio frequency perturbations. The shielded cable should have low coupling capacitance to avoid the possibility of crosstalk between the SDA and SCL lines.

We will discuss bi-directional differential pair solutions, later with bus extension modes.

# Part 3
# The I²C Bus in MultiMaster mode

The proliferation of I²C interfaces makes multi-master mode easier to use. As with all totally asynchronous multi-master systems, the most frequent problems are bus conflicts caused by the collision of messages and propagation times. The I²C bus has a complete multi-master protocol that enables you to engineer local area network (LAN) applications. For this reason, we will discuss multi-user mode in detail.

## Synchronization and Arbitration

### Synchronization

All masters generate their own clocks on the SCL line in order to transmit messages on the I²C bus. The data are only valid during the clock's low period. You need a well-defined clock to enable a bit-by-bit arbitration procedure to take place.

The synchronization mechanism uses the wired **AND** feature of the I²C interfaces' SCL lines. A transition on the SCL line from high to low forces the circuits to start the clock's low period, and maintains the SCL line in low as long as the high state of the clock is absent (Figure 3.19). However, the transition towards the high state of this clock does not change the state of the SCL line if another clock is still in the low period. The circuit that has the longest clock low period maintains the SCL line in the low state.

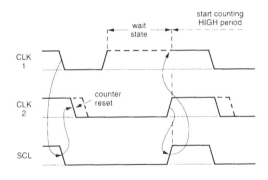

**Figure 3.19**

Circuits with a shorter clock low period go into a wait state at the high level.

When all the involved circuits have terminated their low period, the clock line is released and returns to the high state. At this time, there is no difference between the circuits' and SCL line's clocks, enabling all of the circuits to begin their high period. The first circuit to end its high period pulls the SCL line to the low state, generating a synchronous clock.

*This clock's period is determined by the circuit that has the longest low period. Its high state is determined by the circuit with the shortest high period.*

## Arbitration

Arbitration appears on the SDA line when one master is transmitting a high level while another master is transmitting a low level. This blocks data output from the first master because the level on the bus does not correspond to the first master's level.

Arbitration can continue over an unlimited number of bits.

The first comparison is made on the address bits. If the masters are both in the process of addressing the same circuit, arbitration continues with the data. Since both the addresses and the data are used for arbitration on the I²C bus, no information is lost during the process.

A master that loses the arbitration can generate its clock until the end of the byte. If a master loses arbitration during the address phase, the winning master may be in the process of trying to address the losing master. In that case, the loser must immediately set itself to slave mode.

Figure 3.20 shows the arbitration procedure between two masters. Simultaneous bus access by two masters is first recognized when one

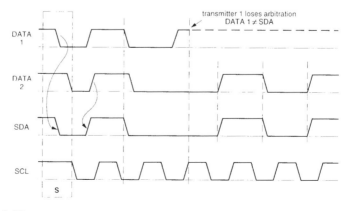

**Figure 3.20**

master drives the bus high while the other drives the bus low. The master driving the bus high at this time suddenly realizes that the level on the bus is not the level that was intended! This in no way affects the data produced by the winning master. At this point, the losing master must withdraw from the bus: the winning master doesn't even realize anything had happened, and continues on it's merry way. In this way, bus control is determined only as a function of addresses and data sent out by the various masters in competition, implying that the bus doesn't need a supervisor or any system of priority management.

## Using the clock synchronization mechanism as handshake

In addition to the arbitration procedure, you can use the clock synchronization mechanism to authorize receivers to 'pause' bus traffic if they need more time to process fast data transfers at the bit or byte level.

### At the byte level

A circuit must be capable of receiving data bytes at a high rate, but may need more time to store a received byte or to prepare another byte for transmission. The slaves can maintain the SCL line at the low level in a type of handshake procedure after a byte is received and acknowledged, forcing the master into a wait state until the slave is ready for the next byte. (See Figure 3.21.)

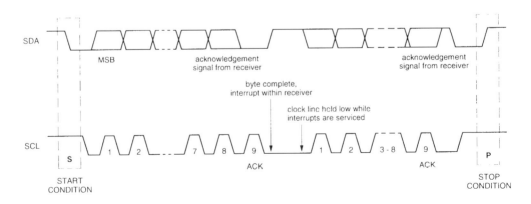

Figure 3.21

## At the bit level

A circuit (such as a Microcontroller without an I²C hardware interface) can decrease the bus clock speed by lengthening each low clock period. In this way, the speed of an I²C system can be adapted to the speed of any master on the bus.

# Part 4

# The I²C Bus, Additional Information

In the I²C bus addressing procedure, the first byte transmitted after the START condition determines which slave will be addressed by the master. The general call address, (which can address all of the circuits), is the exception. When the general call address is used, all circuits are supposed to respond with an acknowledgment. However, you can design some circuits to ignore this address (and many actually are designed this way). The second byte of the general call address defines the action to be taken.

## Addressing

The first seven bits of the first byte define the slave address (Figure 3.22). The eighth bit – the least significant (LSB) – defines the direction of the message. A 'zero' in this position means that the master is going to write information to the appropriate slave. A 'one' means that the master will read information from the slave.

Figure 3.22

When an address is sent, each circuit of the system compares the first seven bits after the START condition with its own address. If a match is found, the circuit considers itself addressed by the master as either a transmitting or receiving slave (as determined by the eighth R/$\overline{\text{W}}$ bit).

A slave address may consist of a fixed part and a programmable part. Because there are usually several identical circuits in a given system, the programmable part of the slave address theoretically determines the maximum number of each type of circuit that can be connected to the bus.

The number of programmable address bits depends on the number of pins available on the package. For example, if a circuit has four fixed address bits and three programmable bits, eight maximum can be connected to the same bus.

The I²C Bus Committee coordinates the allocation of I²C addresses. The address bit combination 1111xxx is reserved for future extensions. The address 1111111 is reserved for address extensions. This implies that the addressing procedure can continue to the following byte or bytes.

Circuits incapable of managing extended addresses do not react to receiving this byte. The seven other possibilities in the 1111 group may be used in future extensions, but are not yet attributed. The combination 0000xxx has been defined as a special group.

## General call address

The general call address can be used to address all of the circuits connected to the I²C bus. However, if a circuit doesn't need information contained in the general call structure, it can ignore this address by not sending an acknowledgment. If a circuit needs data included in the general call structure, it will return an acknowledgment and behave like a slave-receiver. The second byte and those following will be acknowledged by all slaves capable of processing the information. Every slave that is not capable of handling one of these bytes must ignore it by withholding its acknowledgment. The direction of the general call is always specified in the second byte (Figure 3.23).

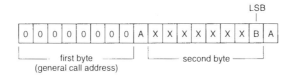

Figure 3.23

There are two cases to be considered:

+ when the least significant bit B is a 0

+ when the least significant bit B is a 1.

## When a B is a 0

The second byte has the following definition (Figure 3.24):

| S | H"00" | A | H"02" | A | ABCD000 | X | A | ABCD001 | X | A | ABCD010 | X | A | P |

Figure 3.24

- 0000 0110 (06h): Software and hardware resets and writes the programmable part of the slave address. When this two-byte sequence is received, all circuits designed to respond to the general call address execute a RESET and accept the programmable part of their address. Precautions must be taken to assure that a circuit does not pull the SCL or SDA line to the low level at power up as these levels will block the bus.

- 0000 0010 (02h): Software only writes the slave address. All circuits that obtain the programmable part of their address by software (and are designed to respond to the general call address) enter into a mode allowing them to be programmed. The circuits do not execute a RESET.

- 0000 0100 (04h): Hardware only writes the address. All circuits whose programmable part of the address is defined by hardware (and which respond to the general call) will latch on to this part on receiving this two-byte sequence. The circuits do not execute a RESET.

- 0000 0000 (00h): This code cannot be used as a second byte. The sequence programming procedure is published in the technical note for each circuit. The remaining codes have not been defined – all circuits must ignore them.

## When a B is a 1

The second byte of the sequence is a *hardware general call*, meaning that this sequence is transmitted by a *hardware master* (such as a keyboard monitor), which can be programmed to transmit to a defined slave address. Since a *hardware* master cannot know in advance to which circuit the message will be transferred, it can only generate this *hardware general call*, followed by its own address, identifying itself to the system (Figure 3.25).

The seven remaining bits of the second byte contain the hardware address of the master. This address is recognized by an intelligent circuit, such as a Microcontroller, which accepts the information from the hardware master. If a hardware master can also function as a slave, the slave address is identical to that of the master. In certain systems, an alternative may exist in which the master-transmitter sets itself to the slave receiver state after RESET. In this way, a master configuring a

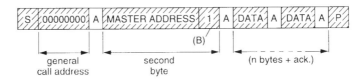

**Figure 3.25**

system can indicate to the hardware master-transmitter (which is now in the slave-receiver mode) to which address the data should go (Figure 3.26). After this programming procedure, the hardware master remains in the master-transmitter mode.

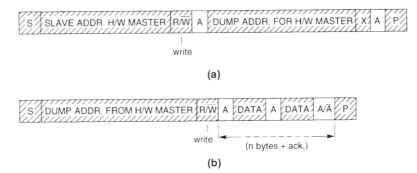

**Figure 3.26**

## Start byte

Microcontrollers can be connected to the I²C bus in two ways. A Microcontroller with a hardware I²C interface on the chip can be programmed to be interrupted only by a request on the bus. When the circuit does not have such an interface, it must constantly monitor the bus by program. Of course, the more time the Microcontroller spends monitoring the bus, the less time it has for its other functions. As a result, there is a speed difference between a fast hardware interface and a Microcontroller that relies on software polling.

In this case, data transfer can be preceded by a start procedure a little longer than normal (Figure 3.27). The start procedure is as follows:

◆ a START condition *S*

◆ a start byte 00000001

◆ an acknowledge clock pulse

◆ a repeated start condition *Sr*.

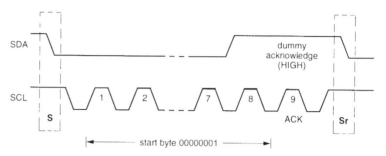

**Figure 3.27**

Once a master has transmitted the S START condition to access the bus, the start byte (00000001) is transmitted. Another Microcontroller can then sample the SDA line at low speed until one of the seven zeros of the start byte is detected. After detection of this low level on the SDA line, the Microcontroller can then switch to a much higher sampling rate to find the repeat START condition $Sr$, which will be used for synchronization.

An acknowledge-related clock pulse is generated after the start byte. This is present only to conform with the byte-handling format used on the bus. None of the devices are allowed to acknowledge the START byte.

# Input/Output I²C Interface Electrical Specifications

The I²C bus allows communication between different kinds of circuits that can also operate with different power supply voltages. For interfaces with a fixed input level, operating at $5\,V \pm 10\%$, the following levels have been defined:

- VIL max $= 1.5\,V$ (low state maximum input level)

- VIH min $= 3.0\,V$ (high state minimum input level).

For interfaces with circuits operating at power supply voltages other than 5 (I2L circuits, for example), these levels must also be 1.5 V or 3.0 V. For CMOS circuits capable of working over a wide range of power supply voltages, the following levels have been defined:

- VIL max $= 0.3\,V_{DD}$ (low state maximum input level)

- VIH min $= 0.7\,V_{DD}$ (high state minimum input level).

The output LOW level for these two groups is:

- VOL max $= 0.4\,V$ at 3 mA

The maximum input current at the low level VOL max of the SCL and SDA pins of an I²C compatible circuit is 10 A. The maximum input current at a high level of 90% $V_{DD}$ of the SCL and SDA pins of an I²C compatible circuit is 10 A, including the leakage current of an output stage (if any). The maximum capacitance of the SCL and SDA pins of an I²C compatible circuit is 10 pF.

Each I²C interface circuit with fixed input levels can have its own power supply of 5 V + 10%. The pull-up resistances can be connected to any of these power supplies (Figure 3.28). However, those I²C interfaces whose input levels are relative to $V_{DD}$ must have a common power supply to which the pull-up resistances are also connected (Figure 3.29).

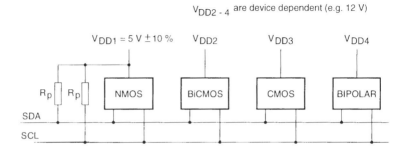

**Figure 3.28**

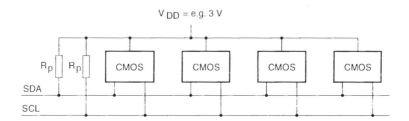

**Figure 3.29**

When circuit interfaces with fixed input levels are mixed with circuit interfaces whose input levels are relative to $V_{DD}$, the latter must be connected to a single power supply of 5 V ± 10% and must have pull-up resistances connected to the SCL and SDA pins (as shown in Figure 3.30).

• The noise margin at the *low* level is 0.1 $V_{DD}$

• The noise margin at the *high* level is 0.2 $V_{DD}$

• Series resistances (Rs) greater than 300 Ω can be used for protection against over-voltages on the SCL and SDA lines (produced by flash-over of a TV tube, for example (Figure 3.31).

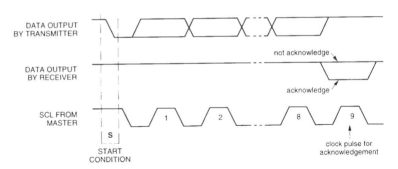

**Figure 3.31**

The maximum capacitance of the bus, is 400 pF. This value includes the capacitance of all wires and pins connected.

# Timing

The clock on an I²C bus has a minimum *low* period of 4.7 microseconds and a minimum *high* period of 4 microseconds. The masters in this mode can generate a bus clock up to a frequency of 100 kHz.

All of the circuits connected to the bus must be able to follow transfers up to a frequency of 100 kHz, either by transmitting or receiving at this speed or by applying the procedure of clock synchronization, forcing the master to enter a wait state and increase the basic period. Naturally, in this case, the transfer frequency is reduced.

# Appendix

You must take special precautions if the arbitration procedure is still under way when a repeated START or STOP condition appears on the I²C bus during a serial transfer.

If this situation can occur, make sure that the masters involved transmit the repeated START or STOP condition at the same place in the bit stream. In other words, arbitration is *not* authorized between:

- a repeated START condition and a data bit
- a STOP condition and a data bit
- a repeated START condition and a STOP condition.

# 4

# The I²C License

## Patents

The beginnings of the I²C bus date back to the mid-1970s. Giving birth to such a big baby wasn't painless! The road to I²C was long; other buses, such as the CBUS (with three wires, clock, data and data line enable), were almost chosen. Philips engineers finally decided to concentrate on a two-wire serial bus.

The protocol and the patents derived from the new serial bus had to take into account all of the parameters and peculiarities of current technologies and new technologies in development (at the time), such as CMOS. Philips filed the first I²C patent under Dutch patent #80 05 976 in Eindhoven, The Netherlands, in 1980. However, a patent must do more than cover a protocol on paper. The text Philips filed covered the characteristics that permit hardware management of this protocol. Anyone who wants to make concrete use of this protocol must sooner or later use circuits described in the patent.

Patents are meant to be used, of course. However, widespread use and exploitation implies sales of licenses, collecting royalties, enforcing law suits for unauthorized use etc.

## The License

An entire judicial arsenal surrounds the document licensing for the I²C bus. As an end-user, you only need to know that this document (and the financial consequences) affects only component or integrated circuit manufacturers (including ASIC houses).

Most system designers need not lose any sleep over it! As a point of information, here is a brief summary of how license regulations may affect component or IC manufacturers:

♦ licensed I²C manufacturers must conform to official I²C bus specifications published by Philips

♦ the components' addresses are assigned by the I²C Committee

♦ licensed I.C. manufacturers have the right to use the official I²C logo for product recognition (Figure 4.1)

♦ In addition to a licensing fee, I.C. manufacturers pay pro rata royalties based on their sales of I²C components, or risk paying a fixed forfeit.

**Figure 4.1**

## License holders

Have you ever wondered who the I²C license holders really are? This question reveals subtle issues:

♦ how popular and accepted is the I²C bus?

♦ who are the potential second sources?

♦ how can we avoid paying license fees?

Here are a few answers.

To our knowledge, as of January 1, 1994, the list of license holders included 32 manufacturers, of whom 20 rank among the 25 leading semiconductor component manufacturers in the world – *source, Data Quest* (see Table 4.1).

Manufacturers who purchase this kind of license don't pay lots of money to frame it and hang it in the living room!

**Table 4.1**

| Ranking | Manufacturer |
|---------|--------------|
| 1 | NEC |
| 2 | Intel |
| 3 | Toshiba |
| 4 | Hitachi |
| 5 | Motorola |
| 6 | Fujitsu |
| 7 | Texas Instruments |
| 8 | Mitsubishi |
| 9 | National Semiconductor |
| 10 | Matsushita |
| 11 | Philips |
| 12 | Samsung |
| 13 | Advanced Micro Devices |
| 14 | SGS Thomson |
| 15 | Sharp |

Most of these manufacturers make components with very specific applications, using their own techniques, for their own specific functions. Others prefer to remain with a line of components that perform general purpose functions. To offer new characteristics or performance, manufacturers have also created derivatives of existing product families (notably in the case of some video components). Others have chosen to simply play the hard and fast second source game. In these cases, it is necessary to study the catalogues of the various manufacturers for each circuit type. It is up to everyone to make the most of this free, competitive market. With the wide variety of $I^2C$ compatible devices covering a multitude of functions, it is the end-user who is ultimately reaps the benefits!

Remaining are the more touchy cases of $I^2C$ buses realized by software in a conventional Microcontroller and those ASIC's or FULL CUSTOM circuits that include a hardware $I^2C$ interface.

# IC Buses in Software: the Problem of ASIC's and Full Custom IC's

The license is very specific:

*Any integrated circuit which can be interfaced to the $I^2C$ bus is considered to be a **product** under the terms of this license.*

You don't get off without a license!

But, not to worry. Except in very unusual circumstances, there is usually no problem when designing with I²C compatible devices– particularly since the world's leading manufacturers and IC foundries are already licensed. A licensed manufacturer should be happy to show you a copy of their license. If not – watch out!

For system designers, it is clearly stated in I²C device datasheets:

*Purchase of Philips I²C components conveys a license under the Philips I²C patent to use the components in the I²C system provided the system conforms to the I²C specification defined by Philips.*

Buy a single I²C component and you are already licensed to use I²C!

## Those not Officially Licensed

Let's consider a few characteristic examples of those not officially licensed:

+ Those who are in full agreement with the protocol . . . except for a few details hidden that don't show up 99% of the time. (It's up to you not to fall into the crack created by this last 1%.)

+ Those who have created their own three-wire buses with the output pins arranged in such a way that a drop of solder can make a short circuit between two, producing a wired AND (entirely by accident), The result, of course, is the I²C bus on the two remaining wires!

It's weird what a little drop of solder can sometimes do! It may necessitate moving from 8 pin DIL or SO packages to 9 pin packages which, of course, do not exist, so the packaging becomes more expensive. In this case, it is the end user who ulitmately violates the I²C license by soldering those two wires together!

The others who are not licensed at all but use the bus anyway, are, by definition **outlaws** and liable to prosecution and/or discontinuation of the of the product.

## Summary

In general, anyone who manufactures components with I²C-bus is required to have a standard I²C-bus license agreement. There are several examples:

+ A company designs an IC (e.g. ASIC) themselves, but has it made by a licensee (licensed manufacturer). In this case no additional license is required.

- ◆ A company designs an IC themself and then approaches a non-licensee for manufacturing. In this case a licence must first be obtained by the manufacturer.

- ◆ A company decides to design and manufacture the IC themself. In this case a license must be obtained.

As far as we know, the terms of the license agreement are today:

1. Payment of 2% royalty on the net sales price of the IC to a maximum of USD 0.35 per IC.

2. Up-front payment of USD 100 000 – 50% of which may be reduced by 2% royalties.

3. Payment of USD 5000 allocation fee for each new slave address.

We hope this very short chapter has cleared up a few obscure points before we proceed to I²C components and their applications.

# THE I$^2$C COMPONENTS

# 5 Structure of an I²C Component

Now that we've discussed theory, philosophy, history, and legality it's time for more concrete subjects. First, we'll describe I²C's structure and the large generic families of I²C components currently in use. Next, we'll analyse specific applications.

The variety of I.C. technologies as well as the various manufacturers who distribute them causes several problems you'll need to consider.

## The Problems of Some Manufacturers and Technologies

There are many kinds of I²C integrated circuits with many different functions. To satisfy these functions, many component manufacturers have had to develop varied technologies.

Bipolar, NMOS and then CMOS, xMOS, ECL, and I²L have all appeared on the market, each with its advantages, its very specific electrical levels and, of course, multiple variations according to the manufacturer. Some fool-proofing has been installed in I²C's protocol, especially relating to the bus data rate as well as the electrical levels, but you can't run a bus on a paper protocol!

Fortunately, one of the principal strengths of the I²C protocol is the ease with which each of the integrated circuit manufacturers can avoid most of the issues created by the technological variety. Now, let's examine the general architecture of a component compatible with the I²C bus.

## General Architecture of a Circuit Compatible with the I²C Bus

Figure 5.1 shows the major features of an 'I²C' circuit's block diagram. It is made up of two principal parts. The **useful** or **functional** part of the

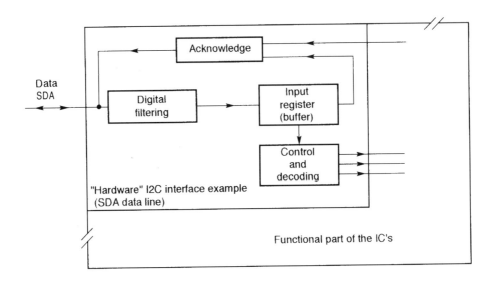

**Figure 5.1**

integrated circuit, i.e. the physical part of the IC where the electrical function that the integrated circuit is designed to accomplish takes place.

Often the electrical qualities of this function determine which technology is chosen for the IC (for example CMOS for low power, BiCMOS for mixed signal, QUBIC for high frequency, etc). However all of the various I²C interfaces must show the same electrical characteristics regardless of which technology is employed.

# The I²C connection interface

The I²C interface's design must integrally satisfy both the electrical and timing characteristics of the bus protocol. The complexity, and therefore physical size of the interface, is directly related to the level of complexity of the function the device is designed to carry out.

Let's consider a few examples:

• The most commonly used interface is also the simplist. Many I²C devices are simple slaves that only receive information coming from a master (for example, the control circuit for a seven segment LED display, or an LCD display driver). In this case, the manufacturer of the integrated circuit can reduce the complexity of the I²C hardware interface and thereby reduce cost. A full-blown I²C interface is not needed.

+ Many devices require the ability to have data written to, as well as read from (for example a memory must have both receiver and transmitter capabilities). This case is somewhat more complicated than the previous, but only requires that the device have slave functionality. No bus Master functions are necessary.

+ If the functional part requires control of the bus in time slices (for instance in a microcontroller), the interface must be capable of managing priorities and bus conflicts. Obviously, this makes the hardware part of the interface significantly larger.

An overall statistical estimate (all applications included) shows that in 80% of the cases, a simple interface is sufficient. Let's examine in a bit more detail the contents and internal organization of this interface to understand its operation and intrinsic qualities.

Serial transmission usually means including a serial entry input register to accumulate the data streaming in. In order to let the sender know that its message has been received (in any case, that its **envelope** has been received) an acknowledgment is also required. To buffer incoming data, an intermediate buffer input stage (made up of a **latch** register) is also provided. The contents of this buffer register are subsequently presented in **parallel** to the part of the circuit that controls the various useful functions of the component after decoding.

Note in Figure 5.1 that the arrow is bi-directional – these buffer stages must sometimes assure real time applications that require acknowledgments returned to the master in indirect time relation with the data received ( for writing data into EEPROMs, for example). There are often either analog (low pass) filters or digital filters of a temporal nature (often employing counting) on both wires (clock and data) of the bus to kill any possible parasites upstream of these registers. An example of the configuration of such an I²C interface is shown in Figure 5.2.

As you can see on the timing diagram in Figure 5.3, you must create an internal clock appropriate to the system (generally ten to twenty times the frequency of the maximum value of that of the I²C). By making use of flip-flops operating in synchronous mode (changing at the rhythm of the clock, after the edges of the incident signal have changed and not on the levels) we can manage to reconstitute a signal at output (D) with the help of gate (C). The signal's width may be different from the incident signal's, but it is free of the parasites that may have been introduced on the SDA data line or on the SCL clock line that could have rendered it totally useless. Obviously, the size of the rejected parasite depends on the value of the internal clock frequency.

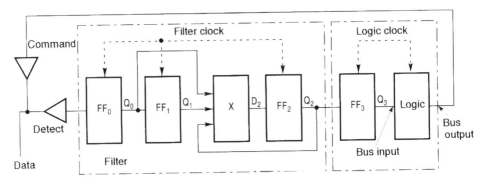

**Figure 5.2**

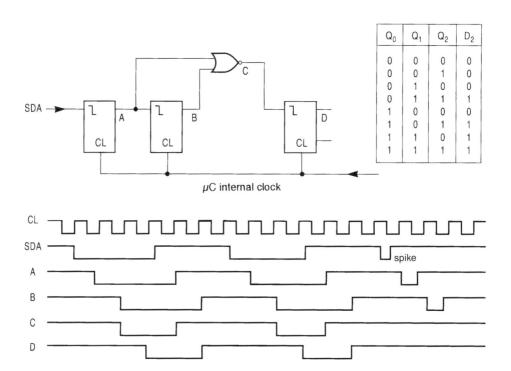

| $Q_0$ | $Q_1$ | $Q_2$ | $D_2$ |
|---|---|---|---|
| 0 | 0 | 0 | 0 |
| 0 | 0 | 1 | 0 |
| 0 | 1 | 0 | 0 |
| 0 | 1 | 1 | 1 |
| 1 | 0 | 0 | 0 |
| 1 | 0 | 1 | 1 |
| 1 | 1 | 0 | 1 |
| 1 | 1 | 1 | 1 |

**Figure 5.3**

In order to give you some order of magnitude, the inteface's integration is accomplished via the equivalent of approximately 200 logic gates in the case of standard slaves, or about 800 gates in the case of complex multi-master applications.

## Other problems

### Translators

In certain types of applications, it is necessary to have electrical level **translators** to align the internal circuit voltages with the more or less floating voltages that may be present on the bus.

### Diode protection

We will discuss the physical aspect of the output stages, as well as cases generated by protection considerations (with diodes) of these stages, in the chapter on protocol.

### Power failure

A problem can arises when you want **hardware** information on whether or not there had been a power supply outage while you were not dealing with the circuit in question.

Now that you know about the **hardware** part of the I²C interface, we are going to present some ideas concerning the existing families of bus compatible components that can be connected. These circuits are known as CLIPS because they can be clipped onto the bus in a modular fashion (even during bus operation in certain software processing conditions).

# The Various Families of I²C Components

The most extensive I²C family involves mass market applications and products, such as car radios and television, but it is also becoming increasingly popular in industrial applications such as telephones, telecommunications, and automotive dashboards.

Let's look at an example of the synergy which has developed in the family of I²C components. Many industrial systems, such as automation and CPU's, require relatively recent, inexpensive components to retain initialization (certain parameters associated with their applications). They make daily make use of I²C EEPROM's, which were introduced to the market a few years ago, mostly for programming mass market video cassette recorders!

Examples of the generic families of existing circuits for mass market applications follow:

- **audio:**
  - tone control
  - volume control
  - A/D and D/A converters
  - source selectors
  - amplifier control
  - fading control
  - all of the specialized circuits for Compact Disc readers

- **radio:**
  - PLL
  - frequency synthesizers
  - IF circuits
  - stereo decoders
  - RDS

- **television:**
  - PAL/SECAM/NTSC decoders
  - TV reception circuits (frequency synthesizers, IF video processing circuits)
  - fast A/D and D/A converters (and there are many more);

and for professional applications:

- **telephone:**
  - amplified reception
  - tone control
  - volume control
  - A/D and D/A converters

- amplifier control

- DTMF generator

- melody generators

- echo canceller.

- **general use:**

  - PLL's

  - frequency synthesizers

  - fast A/D and D/A converters

  - memories (RAM, E2PROM)

  - input/output expanders

  - LED display control

  - real time clocks

  - LCD drivers.

All these applications, and we haven't even mentioned microcontrollers and electronic credit cards! As you can see, these components cover a very wide spectrum.

# 6 I²C Processors and Microcontrollers

## The Possibilities Available

Now you know all about the operation of the I²C bus. All you have to do is to make it work. To succeed, you need the proper components to manage it. There has to be a **master** somewhere in the system!

The **master** function is usually performed by microcontrollers. The purpose of this chapter is to help you discover the possibilities for accomplishing these functions. Many options exist for generating master/slave and transmitter/receiver functions of the I²C.

## Software emulation of I²C

The software emulation, or 'bit-banging', of I²C often makes sense when you're using standard (4/8 bit) or very specific (top of the line) microcontrollers that do not include a dedicated I²C hardware interface. In many cases, applications require only a few I²C functions: E2PROM, clock, RAM, etc. which are simple slaves functions (slave/transmitters of slave/receivers).

An example I²C software emulation using an 80 C51 microcontroller is shown in Figure 6.1. We emphasize the fact that it is always necessary to respect the timing aspects of the protocol (it is up to you to count your cycle times, clock pulses, etc.) based on the speed your micro is running at.

The software written to accomplish this is not very complicated and doesn't require many bytes of code (about 200 to 500 bytes). The software does have two well known and characteristic drawbacks:

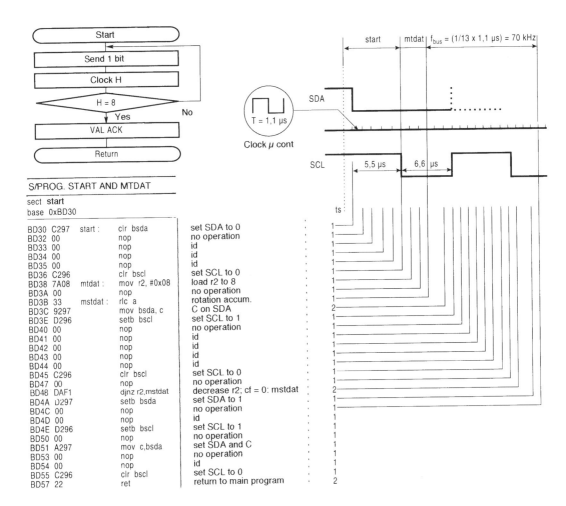

Figure 6.1

• generally, these programs manage I²C by **interrupting** the main program. When the microcontroller CPU is busy with the I²C, **it doesn't do anything else**, which can affect the speed of the entire program if there are frequent accesses to I²C components!

• it is unrealistic from a practical point of view to want to design an I²C **multi-master** using only software emulation. In this case, the microcontroller can spend a significant amount of its time managing bus conflicts or arbitration.

It is sometimes easier to offload I²C overhead to a dedicated I²C interface controller, such as Philips PCF 8584 I²C to parallel bus interface IC (*see*

*Chapter 12)* or to dedicate a microcontroller to this function. Better still, use microcontrollers with dedicated I²C hardware. (For your information, Philips has a complete I²C multi-master software operating with a standard 80 C51, which illustrates the complexity of running a multimaster I²C with software alone).

## Hardware

Most of the I²C licensed component manufacturers offer integrated circuits for hardware management of the I²C protocol for sale. Since no two designers see things exactly the same way, different variants and solutions have appeared, all of which are scrupulously in agreement with the official protocol. Let's consider a few significant examples drawn directly from a designer's catalog to impress you with all of the variations that can be generated by a single manufacturer, all for the same family of microcontrollers.

A basic family (let's take the case of the 80 C51) contains many derivatives – small, medium and large, expensive and economical, with tiny or enormous chips!

Let's return to our I²C **hardware** interface.

♦ **Small**, inexpensive derivative of the family, made on a small chip, in a small package and with the smallest possible I²C **hardware** interface – for example, one managed **bit by bit**. This is exemplified by 8x C751.

♦ **Standard**, inexpensive derivative with a respectable I²C management, **byte by byte**. That is the case for microcontrollers of the 84 Cxxx, PCD 33xx, 90 CLxx families.

♦ **Large** derivative of the family, on a large chip in a large package, with the possibility of a dimensionally plump I²C hardware interface with intelligent, **byte by byte** management – for example, the 8x C652, 654, 552.

♦ **More complex** derivative of the family on an enormous chip and the largest package economically possible, with little room remaining to insert a hardware I²C interface and, unfortunately, a more evolved, bit by bit architecture – for example, the 8x C528.

These simple examples show you how difficult it is to try to maintain the same type of hardware interface for all of the members of a single family.

Despite this hardware diversity, the user intent on software compatibility has not been completely forgotten.

## Hardware implications of the software

Fortunately, there are sophisticated professional software tools, such as **Cross Assemblers** and **Cross Compilers** (PL/M or C), to help resolve this delicate and painful problem. These tools easily allow the personalization of most of the various types of microcontrollers existing on the market (with the help of preliminary declarations of **define, set, declare, include**).

The fact that you can **specify** particular characteristics and specific resources, allows you to construct routines for the many I²C hardware configurations that seem to be transparent or independent of the microcontroller selected.

These programs make use of elementary subroutines that always have the same names. For example I²C_start, I²C_stop, I²C_number of bytes transmitted. This allows processing the I²C call routines of various types of microcontrollers as software modules pompously baptized **universal**, frequently used with the UART, A/D and D/A converters, etc.

But, every universal software rose has its hardware thorn. Although each module may be optimized with respect to the number of bytes generated, each also has a particular overhead reflecting the complexity of the I²C hardware of the specific microcontroller. It is always worth while examining the length of the code generated for each of them to avoid excessive demands on the program memory size.

## Examples of Various Circuit Configurations of I²C Hardware

We have selected configurations that are particularly interesting because of their efficient management of the I²C bus.

This entry into circuit design practically forces us to ignore hardware solutions operating on the bit level. Although powerful, they almost always require the microcontroller to function in interrupt mode and are better adapted to specific tasks which allow this mode of operation. (Examples of applications: decoding remote control commands, keyboard controller, slave-only applications, etc.)

We present two classical architectures representative of many systems; the standard **byte by byte** types of 8 bit 84 Cxxx, PCD 33xx families and the 16 bit 90 C1xx families and, the **high performance byte by byte** type associated with 8x C652, 654, 552 and 16 bit (at the present level of development) 90 C2xx and XA microcontrollers.

# Byte Oriented I²C Hardware Interface

This section will describe the 8-bit microcontrollers, for example, 84 Cxxx, PCD 33xx and 16 bit microcontrollers of the 90 C1xx type.

Figure 6.2 shows the block diagram of the serial interface. The serial bus SCL clock line uses pin 3, which is exclusively reserved for it, and the SDA data line uses pin 2.

The exchanges between the I/O serial interface of the CPU are made by the internal bus, under the control of an interrupt request line. Four registers are used to store the data and information controlling the operation of the interface. They are:

+ the data shift register (S0)

+ the status word of the serial interface (S1)

+ the clock control word (S2)

+ and, finally, the address register.

# Shift Register S0

This register is used for the series-parallel conversion of data. The data to be transmitted is loaded into S0 by the microcontroller, then shifted bit by bit towards the output. The most significant bit is transmitted first. The data received by the serial bus are entered, bit by bit, into S0, the most significant bit first. After transmission of a complete byte, or reception of a data byte, a specific address byte or the general call address, an interrupt is generated.

## *Status word S1*

This register gives the microcontroller information on the status of the interface and also stores the control bits loaded by the microcontroller that determine the operating mode of the interface.

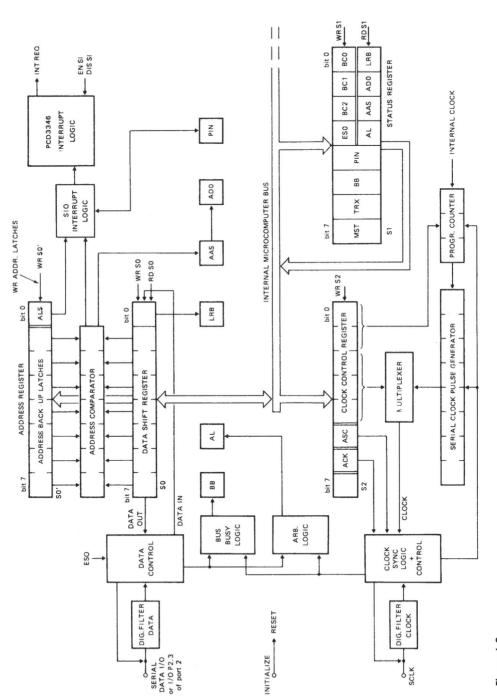

Figure 6.2

The setting of the various bits of S1 is shown in Figure 6.3. Notice that bits 0 to 3 are doubled: the control bits that occupy these positions can only be written by the microcontroller, while the bits of the interface status can only be read.

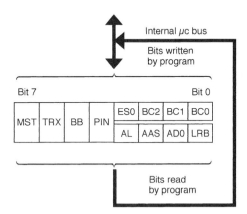

**Figure 6.3**

The **MST and TRX** bits, 7 and 6, determine the operating mode of the device.

- MST (Master) controls the operation of the interface in master or slave configuration (MST = 1: master device, MST = 0: slave device).

- TRX determines the direction of the exchange (TRX = 1: data transmitter, TRX = 0: data receiver).

The four modes that result from the possible combinations of MST and TRX are summarized in the following list:

- **Slave-receiver:** a device which receives data and clock signals from a master-transmitter.

- **Slave-transmitter:** a device which has been addressed by a master-transmitter requesting data transmission; the slave-transmitter sends the data, synchronized with the clock signal furnished by the master-receiver (the master-transmitter previously configured itself as a master-receiver in order to receive the data from the slave-transmitter).

• **Master-receiver:** a master who, after initializing a read operation, receives data from a slave transmitter. The master receiver always generates the clock.

• **Master-transmitter:** all bus masters begin as master transmitters. The value of the R/$\overline{W}$ bit then determines its next role (i.e. master-transmitter or master-receiver).

| MST | TRX | MODE |
|-----|-----|------|
| 0 | 0 | slave-receiver |
| 1 | 0 | master-receiver |
| 0 | 1 | slave-transmitter |
| 1 | 1 | master-transmitter |

The **bit BB** (Bus Busy) indicates the status of the bus. Each time that a start message is detected on the bus, the bit BB is set. It is then reset to zero by the detection of an end of message. If a master-transmitter tries to generate a message start while the bit BB is set, the indicator AL (Arbitration Lost) will be set in turn and an interrupt generated. The start of message will not be transmitted.

The **bit PIN** (Pending Interrupt Not) set to 1 signifies the absence of a pending interrupt; PIN = 0 indicates the presence of a pending interrupt, which allows an interrupt request to come from the serial interface (if it has been authorized). The pending interrupt is reset (PIN = 1) each time data are read from or written to the S0 register. An interrupt is generated in one of the four following cases: a complete byte has been transmitted; the bus is lost after arbitration; a complete byte has been received or the correct address of the circuit has been detected.

The **bit ESO** (Enable Serial Output) validates or invalidates the serial interface. When ESO = 1, the serial interface is validated and pin 2 becomes the data line. When ESO = 0, the serial interface is invalidated and pin 2 is the parallel input/output P23; the identification address register can then be loaded by writing into S0. Finally, it should be noted that the bit ESO can be set by software, but cannot be read.

The bits **BC0, BC1,** and **BC2** (Bit Count) make up a counter that indicates the number of bits in the word to be received or transmitted. This counter can be set, but not read, by the CPU. The contents of the counter directly show the number of bits per word, with the exception of the value 0, which indicates a word of 8 bits.

The **bit AL** (Arbitration Lost) signifies loss of access to the bus after arbitration; this indicator is set by the logic when the serial interface loses access to the bus after the arbitration of a conflict.

A pending interrupt will always be generated after reception of the word during which access to the bus has been lost. When the CPU decides to read the status word, the bit AL indicates that the transmission requested has not taken place. The bit AL is reset to zero at the same time as the pending interrupt is erased (PIN = 1).

The **bit ASS** (Addressed As Slave) is set by the logic when the interface detects either its own specific address or the general call address in the first byte of the message (if it has been programmed to function in the address recognition mode). This bit ASS is reset to zero at the same time that the pending interrupt is erased.

The **bit AD0** (Address zero) is an indicator set by the logic after detecting a general call, when the interface operates in address recognition mode. The general call address consists of a byte containing eight zeros. The bit AD0 is reset to zero when the bus becomes free after a stop message.

The **bit LRB** (Last Received Bit) is the last bit received. This bit contains either the last bit of the received data or the acknowledgment signal coming from a receiver, for a transmitter operating in the 'with acknowledgment' mode.

## Clock control register S2

The five least significant bits of S2 are used to program the serial clock counter. The serial clock signal is derived from the internal clock of the microcontroller, which is equal to one third of the frequency generated by the quartz crystal oscillator.

With a 4.43 MHz crystal, the frequency of the serial clock can be programmed in the range from 720 Hz to 100 kHz, using two dividers $m$ and $n$. As shown in Figure 6.4, the bits 0 and 1 define the value of the divisor $m$ and the bits 2, 3 and 4 specify the value of the divisor $n$. The frequency of the serial clock is then given by the equation:

$$f(\text{clock}) = f(\text{crystal})/3m \times n$$

The **bit ACK** (Acknowledge) is reserved for acknowledgment. For the acknowledgment procedure to take place, the transmitter and the receiver must both have the indicator ACK = 1. The master device will then transmit an extra clock pulse after each complete word, as shown in Figure 6.5. During the additional clock pulse, the transmitter maintains the data line in the high state and the receiver acknowledges the message by setting it to the low state. The transmitter stores the received acknowledgment signal in the bit LRB of register S1.

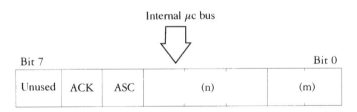

Internal μc bus

| Bit 7 | | | | Bit 0 |
|---|---|---|---|---|
| Unused | ACK | ASC | (n) | (m) |

| Register S2 | | | n |
|---|---|---|---|
| bit 4 | bit 3 | bit2 | |
| 0 | 0 | 0 | 2 |
| 0 | 0 | 1 | 4 |
| 0 | 1 | 1 | 8 |
| 0 | 1 | 1 | 16 |
| 1 | 0 | 0 | 32 |
| 1 | 0 | 1 | 64 |
| 1 | 1 | 0 | 128 |
| 1 | 1 | 1 | 256 |

| Register S2 | | m |
|---|---|---|
| bit 1 | bit 0 | |
| 0 | 0 | 5 |
| 0 | 1 | 6 |
| 1 | 0 | 7 |
| 1 | 1 | 8 |

**Figure 6.4**

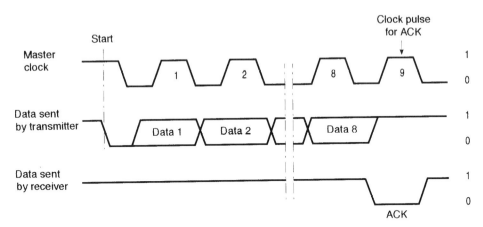

**Figure 6.5**

The **bit ASC** (Asymmetrical Clock) signifies that the clock is asymmetrical. If ASC = 1, the clock generator provides a signal in which the duty cycle ratio low/high is 1/3. In this case, the divisor $n$ must be 128 (that is to say that the bits 4, 3 and 2 of S2 are, respectively, 1, 1 and 0). The frequency of the clock is then given by:

$$f(\text{clock}) = f(\text{crystal})/384m$$

## Address register

The address register (Figure 6.6) contains seven latch flip-flops for the address and the indicator ALS (Always Selected). The address flip-flops contain the specific code of the device, while the bit ALS determines the circuit response mode: if ALS = 1, the serial interface will respond to all messages, whatever the address received; on the other hand, if ALS = 0: the serial interface responds only to the messages containing its own address or the general call address.

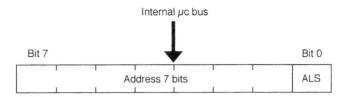

**Figure 6.6**

In this last operating mode, the least significant bit of the first word received determines the direction of the subsequent exchange (1 for reading, 0 for writing). In addition, when a device detects its own address or the general call address, it automatically sets the bit TRX of the status word to the value indicating the direction of the exchange requested (reception or transmission). The general call address (eight zeros) corresponds to the address 0 and the read/write command is, thus, zero (writing).

The address register can only be loaded by the microcontroller's internal bus when the serial interface is unused (ESO = 0). The instruction MOV S0,A is then used, which will transfer the contents of the accumulator to the address register instead of to S0, as is normally the case.

## Interrupt logic

The interrupt mechanism can be validated by the instruction EN SI, or inhibited by the instruction DIS SI. When interrupt logic is validated, the generation of a pending interrupt is translated into an interrupt of the serial interface towards the CPU.

To service that interrupt, the processor will execute the instruction to be found at address 5 of the program memory (vector). When the interrupt logic is inhibited, it is always possible to service a pending interrupt by polling the status word of the serial interface, by the processor, and reading of the signal PIN.

During a data transfer sequence, the first pending interrupt is produced in one of the five following cases: a complete word (data or address) has been sent; the general call address has been received; the identification address of a circuit has been received; eight bits have been received and ALS = 1; access to the bus has been lost after arbitration and a complete word has been received. When, for one of these reasons, the first interrupt has been installed, the other interrupts are normally processed after each complete word.

As of the moment when a pending interrupt is generated, the clock line is maintained in the low state until the interrupt is serviced. That implies a certain delay between the words transmitted, which will depend on the execution time of the interrupt procedure by the devices taking part in the exchange. It is only when the clock line is liberated that the transfer of the following word can begin.

## Operation of the interface

From the moment when a device generates a start message, all of the BB (bus busy) indicators in the system are set and no other circuit can initialize a transfer until the first has freed the bus. It is possible for two or more devices to transmit a start message practically at the same time, so that none of them can detect the start messages of the others. Then it becomes necessary to use an arbitration procedure that can give control to a single circuit and provide for the synchronization of the clocks during the arbitration.

# High-Performance, Byte-Oriented I²C Hardware Interface

Examples of 8-bit microcontrollers include the 80 C552, 652, 654 interfaces and 16-bit processors 90 C2xx.

Our goal is to show you where and how to arrange the sub-routine that represents the I²C interface program of this component **intelligently**. First, we need to describe the make-up of this I²C (hardware) interface in detail. (We ask software lovers to wait; we'll get back to you soon).

The I²C interface hardware of the 80 C552 covers a significant area of this component's chip (see Figure 6.7) because it was designed to satisfy all of the I²C specifications (master, slave, transmitter, receiver, in mono- or multi-master mode).

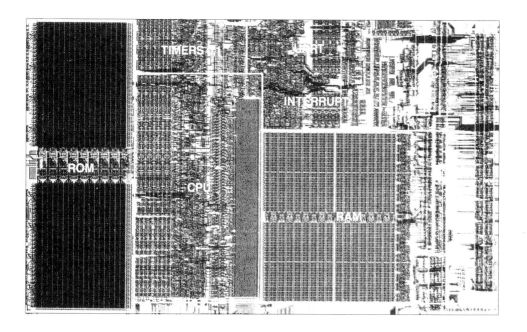

**Figure 6.7**

**Note:** Everything in this paragraph applies directly to the 80, 83, 87 C552, C652 and C654, which have a strictly identical I²C hardware interface.

In academic language, this interface is called SIO1 (Serial Input/Output 1, the 0 being for UART), whose flow diagram (Figure 6.8), shows port P1.6 for the SCL clock and port P1.7 for the SDA data. These ports must normally be at rest, in the high logic 1 state (in the absence of all other commands).

This SIO1 interface can be controlled entirely by software, with the help of four SFR registers, whose names and functions follow:

| Register | Description | at the SFR addresses |
|----------|-------------|----------------------|
| S1CON | S1 CONTROL | D8h |
| S1STA | S1 STATUS | D9h |
| S1DAT | S1 DATA | DAh |
| S1ADR | S1 ADDRESS | DBh |

- S1CON manages the interface

- S1STA is a spy that reports what is going on the bus

- S1DAT is a data mail box in both transmission and reception

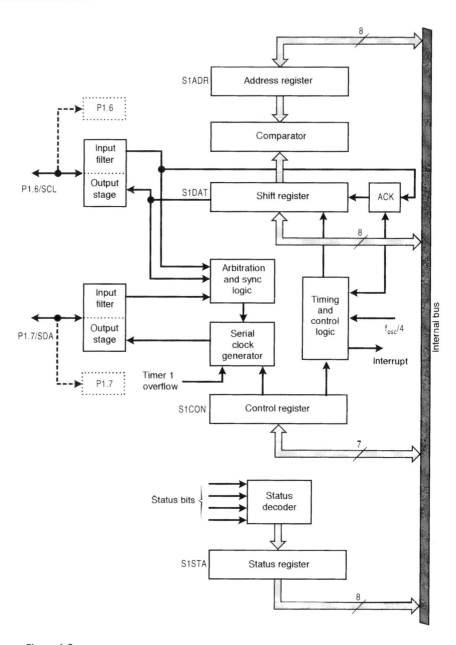

**Figure 6.8**

- S1ADR   represents the I²C address (modifiable, if necessary) of the 80C 552 component.

Let's look at some of the electronic elements included in the SIO1 interface. It is made up of:

### Input filters

The filters are internally located on the microcontroller's SCL and SDA pins. They are **digital** in design, and are sampled at a quarter of the crystal clock frequency of the microcontroller, rejecting all parasites shorter than three oscillator periods.

### Output stages

Despite the fact that the electrical levels are in agreement with the I²C bus specifications, the outputs are **real** open drains **without** any protection diodes tied to $V_{DD}$. This makes it possible to cut microcontroller power where several masters live together on the bus without perturbing bus operation (on SCL and SDA).

### SCL generator

Internal dividers by 60, 120, 160, 192, . . . 960 divide the pilot frequency of the microcontroller crystal oscillator to create an SCL signal of programmable frequency, with a constant duty cycle of 50%.

### Arbitration and synchronization logic

Wired logic on the chip accomplishes the delicate job of arbitration and synchronization, freeing us from acrobatic software that gobbles up code and microcontroller CPU time. Arbitration and synchronization logic handles the complex bus operating phases in **multi-master** mode without giving you ulcers. Engineers have reason to thank the manufacturer of this wired chip logic!

### Registers

These four registers are the very soul of the SIO1 interface. Judiciously placed in the SFR's (Special Function Registers) at bit addressable locations, they simplify software operation and take proper advantage of this interface and I²C bus. The best way to really understand how they operate is to study each of the four SFR's.

### S1CON (S1 CONtrol)

This bit-by-bit addressable eight-bit SFR register (Figure 6.8) controls all the S1O1 interfaces's structural functions. Each of its bits has a precise task:

| bit number | 7 | 6 | 5 | 4 | 3 | 2 | 1 | 0 |
|---|---|---|---|---|---|---|---|---|
| S1CON (D8h) | CR2 | ENS1 | STA | STO | SI | AA | CR1 | CR0 |

## CR2 , CR1, CR0 bits

The function of this registers 0, 1 and 7 bits is to define the I²C bus data flow rate, i.e., define the frequency of the SCL clock (when the microcontroller acts as master). The following table gives the relations between the data rate, the microcontroller clock frequencies and the settings of these bits. Whatever the frequency of your crystal oscillator, we suggest that you operate at the maximum I²C bus frequency , (usually, 100 Kbits per second for most components).

| | | | Bit frequency (kHz) at fosc | | | |
|---|---|---|---|---|---|---|
| CR2 | CR1 | CR0 | 6 MHz | 12 MHz | 16 MHz | Fosc divided by |
| 0 | 0 | 0 | 23 | 47 | 63 | 256 |
| 0 | 0 | 1 | 27 | 54 | 71 | 224 |
| 0 | 1 | 0 | 31 | 63 | 88 | 192 |
| 0 | 1 | 1 | 37 | 75 | 100 | 160 |
| 1 | 0 | 0 | 6.25 | 12.5 | 17 | 960 |
| 1 | 0 | 1 | 50 | 100 | 133 | 120 |
| 1 | 1 | 0 | 100 | 200 | 267 | 60 |
| 1 | 1 | 1 | >0.25 | >0.5 | >0.67 | 96x (256 − Reload value Timer 1) |
| | | | <62.5 | <62.5 | <56 | (reload value range: 0–254 in mode 2) |

This does not prevent you from allowing microcontrollers to 'chat' more rapidly with each other over the I²C wires in **multi-master** applications. You can create your own data rate with the help of the last lines of the table (use of timer 1).

Let's consider another bit of the S1CON register:

## The ENS1 bit (ENable S1)

ENS1's task is to enable the SIO1 interface. There are two possible cases:

1. ENS1 = 0: When ENS1 = 0, the internal lines leading SDA and SCL to P1.6 and P1.7 pins are set to a high impedance state; these pins can then be used as conventional open drain I/O ports. (To assure proper operation, I²C requires open drain outputs with external pull-up resistances to +5 V.)

2. ENS1 = 1: In this case, the entire I²C SIO1 interface is activated and ready to operate. The P1.6 and P1.7 ports are normally preset to 1 by

software (using the assembler instruction SETB (set bit), for example), since the P.1x ports are among the SFR's addressable bits. Logically, you open ENS1 = 1 when many other bits are already preset to values of your choice.

## The STA bit (STArt)

STA's function is to start the beginning of the I²C exchange.

1. STA = 1: Let the battle begin. Setting this bit to 1 tells the I²C SIO1 interface that you want to start an exchange as the communication 'boss'. The interface electrically generates a **start** condition on the I²C wires after making sure that they are free. If the wires are occupied, the interface waits politely for a **stop** condition to liberate the bus. If you reset STA = 1 when you are already in the master-transmitter state you issue a second start, or the RE START condition, which is very practical for accessing certain I²C components such as memories (RAM, E2PROM).

2. STA = 0: in this case, don't start or restart! You remain a slave during this communication.

## The STO bit ( STOp)

1. STO = 1: The master (you) wants to terminate the exchange. By setting the STO bit to 1, you tell the interface to electrically generate a **stop** condition on the I²C wires. Since the interface systematically rereads whatever transpires electrically on the bus wires, it detects the **stop** condition and sets STO to 0. If STA and STO are simultaneously equal to 1, a stop is first generated and a new start is created.

2. STO = 0: in this case, don't STOp!

## The SI bit (Serial Interrupt)

1. SI = 1: When the SI bit is set to 1 and the interrupt authorization register's EA and ES1 bits are also set to 1, the I²C SIO1 requests its own interrupt. The SI bit is set to 1 by the microcontroller's internal electronics when it detects one of the 26 possible situations than can occur during transmission. This momentarily suspends operation of the I²C bus. After having deciding what to do next, it resets by software.

2. SI to 0: An exception is the case where the value of S1STA(tus) is F8h. This case indicates that nothing important is going on, there is no interrupt request and SI remains obediently at 0.

3. SI = 0: No interrupt is requested nor intended – the status codes don't trigger anything.

## The AA bit (Assert Acknowledge)

1. AA = 1: If the AA bit is set, an **acknowledge** is returned during the ninth pulse of the SCL clock if:

   ◆ the correct slave address has been received

   ◆ a general call has been received

   ◆ a byte of data has been received in master-receiver mode

   ◆ a byte of data has been received in slave-receiver mode

   ◆ AA = 0: If the AA bit is reset, a **non-acknowledgment** will be returned during the ninth clock pulse. (Non-acknowledgment signifies the voluntary intention not to acknowledge). This happens when:

      ◆ a byte of data has been received in master-receiver mode

      ◆ a byte of data has been received in slave-receiver mode.

For more extensive information, consult the manufacturer's documentation, which includes all of the variants and sub-variants possible on the I²C bus on acknowledgments and intentional non-acknowledgments.

## S1STA (S1 STAtus register)

This register provides continuous information on the progress of the exchange taking place, and any problems that may have occurred during the exchange. Detailed study shows that there are twenty-six (26) incidents that can occur, mobilizing at least 5 register bits to represent them digitally. The constitution of the S1STA shows that the 5 most significant bits of the register have been chosen for this representation and the lower order bits are left at 000.

For this reason, the hexadecimal values of the bytes formed in S1STA can only end with a 0 or an 8 in hexadecimal (example: D0h or C8h). The values obtained are spaced in steps of 8. The very nature of these values provides great flexibility in implementing the control software (see the chapter on software design).

When the SIO1 interface is put into operation, the device starts up and provides continuous information on the progress of the exchange. Figure 6.9 illustrates a detailed example of master-transmitter, one of the four modes. The manufacturer shows all possible eventualities in the specifications, (presented in Section 5).

As soon as you set STA(rt) to 1, the S1STA status register loads itself automatically with the A8h value (see Figure 6.9). This loading will trigger an interrupt. You need to process the interrupt to see if you are in agreement with the contents of the STATUS register and decide whether or not to continue with the rest of the exchange. This process continues throughout the exchange until the final stop.

This process is extremely efficient, assuring the reliability of the exchange.

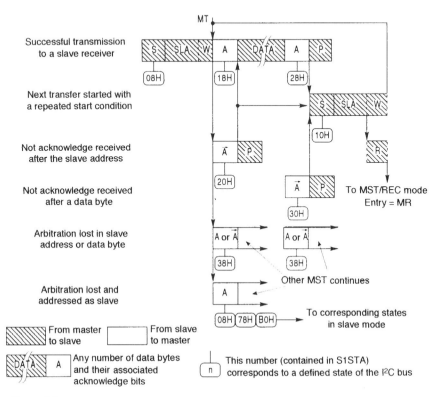

Figure 6.9

## S1ADR (S1 ADdress)

You write the 7-bit name of your microcontroller in this register so that you can communicate with it. It's name is its one-**address** register in slave mode. This register has no meaning when the microcontroller is master. The name must be written in the seven most significant bits of the S1 ADR register byte. The last (lowest order) bit makes it possible to create the value necessary for general call reception. The most significant bit corresponds to the first bit received after the START condition on the I²C bus and a logic 1 corresponds, electrically, to the high level on the bus.

## S1DAT (S1DATa)

This last register contains the data to be transmitted or the data that just arrived. The CPU can either read this register or write into it. The register always contains the last byte that was present on the bus. For more extensive information, see the software in the applications chapter to understand better how to read and write into this register during an exchange.

This chapter includes the basic information you need about the very different I²C hardware interfaces on the market. The next chapter considers applications (control software) for these interfaces and their intelligent installation in the microcontroller's RAM and (EP)ROM data memory.

# EXAMPLES OF APPLICATIONS OF THE I²C BUS

# 7 Design of I²C Microcontroller CPU's

## Introduction

Part Three presents examples of several different I²C hardware 'modules' which can be used in a variety of applications. We will discuss many basic concepts in detail. We have kept the basic hardware and software architecture open to enable you to integrate your own ideas. I²C architecture lends itself to a modular systems approach with many possible configurations. You can customize these examples to fit your own application.

In Part Three, you will:

♦ become familiar with the intimate life of an I²C microcontroller

♦ learn how to design a programmable I²C platform that you can use for a variety of purposes

♦ learn about several off-the-shelf I²C components that are easy to use, inexpensive, and offer a variety of functions

♦ understand the connections and relationships between hardware and software that affect your design.

For clarity, our discussion will be split into two parts: hardware and software.

Our hardware discussion will address:

♦ The I²C-platform hardware and the reasons for choosing a specific design

- the practical uses of a 'I²C controller card'

- description of several I²C modules

- the layout of the principal printed circuits (included at the end of the book).

Our software discussion includes:

- microcontroller architecture and software design choices

- software philosophy for the 8x, C751, 528, 552, 652 microcontrollers

- software in assembly language for the 80 C31, 80 C652 and 80 C552

- how to adapt software to 8052 AH_BASIC

- C high-level language software.

# The Hardware Part

## *The I²C control platform*

The following modules manage most of the functions needed in standard industrial and domestic application environments. We have broken down our discussion in two parts:

1. the programmable I²C platform
2. the peripherals operating under the I²C bus protocol.

Chapter Six discusses the extensive family of functions that you can connect directly to the I²C bus.
The I²C control platform is made up of:

- an 8-bit I²C control platform on a card (PCB) which includes the microcontroller, RAM and ROM memories, and standard components, such as decoding logic LS(HC), 138, LS(HC) 573, etc. (Figures 7.1 and 7.2).

- a serial I²C bus connection that can be interconnected with many different independent modules or functions, for example a remote control or LCD display.

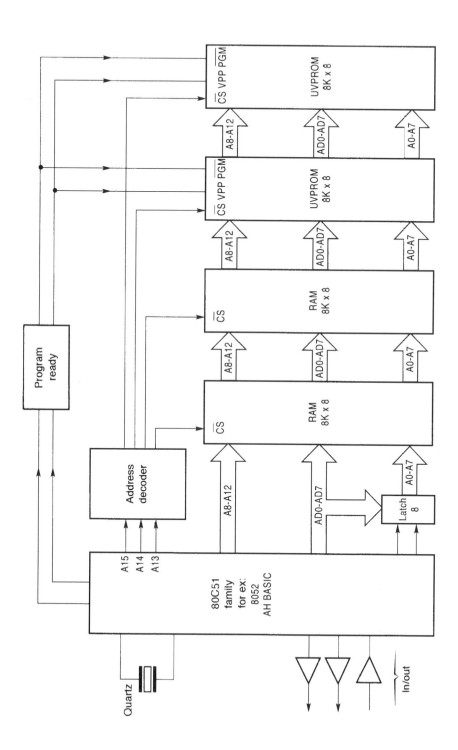

Figure 7.1

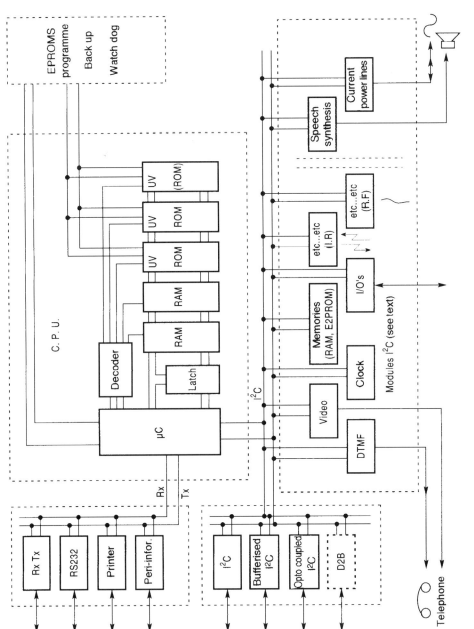

Figure 7.2

+ re-programmable memory that enables you to customize the system to your environment.

This approach has the following advantages:

+ modularity: peripheral modules can be easily connected and disconnected

+ easily adaptable to different transmission media

+ conceptually easy to understand

+ standard mechanical format

+ easily reconfigured.

Before designing your own $I^2C$ platform, **carefully read** the following paragraphs. Paying careful attention will enable you to design your own application efficiently and effectively. Here are some basic guidelines.

# Choice of bus connnections

Although the purpose of this platform is to control the $I^2C$ bus, the 8-bit parallel bus is also available for compatibility with other types of (parallel bus) peripherals (although we will not make use of it in our examples). This will make your $I^2C$ bus platform truly general purpose and usable for many other projects.

# Choice of memory partitioning

We have physically divided the program memory into blocks of 8K × 8 bits to provide modularity and provide the electronics student with an easy-to-understand layout. Of course, the entire memory may be integrated in a single IC. In the commercial world of mass production where low cost and component count are critical, the program memory is usually integrated onto the microcontroller itself, and mask-programmed during chip production.

## Programming the memory

There are many ways to store the program software, such as ROMs, PROMs (OTP), UV EPROMs, E2PROMs, etc. A good choice for our purposes is UV EPROMs because they are easily erased and reprogrammed, and are cheap and readily available.

## Choosing the microcontroller

Your choice of microcontroller plays a big part in your $I^2C$-based design. This is true while the choice of microcontroller often dictates how expensive your design will be (the micro is often the most expensive component), the speed of your system (important in real-time applications, or multi-tasking environments), and even what type of developement tools will be required. The family of microcontrollers you choose should be flexible, powerful, adaptable, easy to use but also economical. And for our purposes we don't want a design that requires costly development tools.

The best choice is an easily available, flexible, well established microcontroller. A good choice is the CMOS derivatives from Intel's 8051 MOS microcontroller family, known as 80 C31, 80 C52 or the Philips' derivatives 80 C652, 80 C654, or 80 C552. The latter include Philips Semiconductors' integrated $I^2C$ hardware interface.

This extended 8051 family offers many interesting functions:

+ a large capacity for addressable external memory (64K of RAM and 64K of ROM)

+ multiple timers

+ $x$K of (EP)PROM internal memory, as needed.

This family also includes a special type with resident program in internal memory that understands and interprets BASIC – the **NMOS 8052 AH BASIC** microcontroller. Although BASIC is a relatively simple language, using it together with the 8052 AH_BASIC micro means you don't need to buy an additional BASIC interpreter. A BASIC $I^2C$ bus routine (hex file) is included on the disk provided with this book.

The biggest benefit is that Philips' CMOS micros 80 C652 and 80 C654 derivatives and Intel's 8051 or 8052 AH BASIC not only have integrated $I^2C$ hardware interface, but are pin-to-pin compatible with each other.

Designers can remove the 8052 AH BASIC after debugging their system and replace it with the microcontroller best adapted to the level of performance that their application requires at a reasonable price. (Considering the price of the 8052 AH BASIC, we recommending holding on to it to use for future debugging).

# Basic card characteristics

## Hardware

The I²C controllers 'brains' consists of the following parts:

### CPU and memory

- Philips' 80 C652 microcontrollers with integrated I²C interfaces, or an 80 C31 or Intel's 8052 AH BASIC

- $(x)$ K × 8 RAM

- $(x)$ K × 8 EPROM or OTP (or EEPROMs as an option)

- misc. logic

### I/O interfaces

- Rx and Tx transmission ports

- a serial RS 232 link

- a peripheral link connector

- a standard I²C output port

### Software

- modular software

- BASIC language option (when using 8052 AH_BASIC)

- examples in C language (given on the diskette supplied with this book)

- simple, open, examples of various I²C peripherals (also on the diskette)

# I²C Function cards and module characteristics

The function card consists of the I²C bus and the modules that connect to it, summarized below.

The I²C controller platform PCB (see end of book) includes a section for connecting I²C modules. This section may be separated (cut) and connected via jumpers or cables. Examples of I²C module PCBs are also included in the back of this book.

### Hardware modules

+ real time clock

+ memories (RAMs, EPROMs)

+ parallel ports for input/output, such as keyboards

+ LED displays (four digits, seven segments)

+ LCD displays (four digits, even segments, twenty or forty characters, 35 points)

+ infrared transmission

+ infrared reception

+ DTMF

These are just a few examples of a variety of functions which can be plugged into the function card and easily controlled via I²C.

### I/O interfaces on the function card

+ standard I²C

+ buffered I²C

+ electrically isolated I²C using an optical coupler

+ D2Bus

+ digitally buffered parallel input/output ports

Other possible I/O interfaces include:

+ telephone line

+ electrical distribution network ('powerline modem')

+ analog input/output

+ IR remote controls

+ sound input/output (buzzers, loud speakers, tone decoders).

# 8

# Electrical Circuit Diagrams

We can divide the I²C's electrical circuit diagrams into two categories:

1. The I²C controller platform (processor plus memory and 'glue' logic).

2. The I²C function modules (peripherals and associated passive components).

In this chapter we will describe the microcontroller-based I²C controller platform.

## Circuit Diagram of the CPU

The schematics illustrated in Figures 8.1 and 8.2 are very conventional for this type of microcontroller. We assume the reader has some background in microprocessor system design, and we will not go into too much detail.

Address demultiplexing is accomplished by the well-known 74HC138. We use the 74HC573 for the address latch. Note that we have brought out all of the necessary lines to address the maximum possible memory field (all address lines A0 to A15 are connected).

The remaining control signals such as READ, WRITE, CHIP ENABLE, CHIP SELECT, OUTPUT ENABLE, are connected via the AND gates, which have been judiciously distributed to assure correct system operation.

At this stage, we have dealt with many of the microcontroller pins. There are still a few of the forty remaining pins that we are going to examine now.

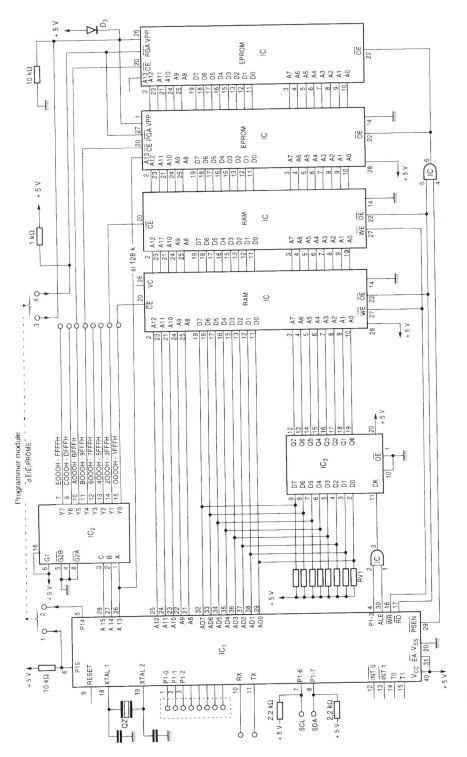

**Figure 8.1**

## The oscillator

The oscillator has two pins for the microcontroller clock. To make it oscillate, use a quartz crystal and two capacitors (the capacitors insure oscillator startup when the power is first switched on).

Be careful when choosing the crystal; too often, a poor quality crystal can cause erratic operation. We suggest that you use standard good quality crystals with a series resistance of approximately $50\,\Omega$, and external capacitors that conform to the value of the load capacitance recommended by the manufacturer. (Our case specifies 27 pF.)

For the crystal frequency, use 11.0592 MHz, of course! Why do we attach 'of course' to this rather strange value? As strange as it may seem, this frequency is standard because of the way in which the 8051 family's internal timers are designed. It is easy to configure one of the internal timers to obtain a baud rate exactly equal to 1200, 2400, 4800 ... bauds when the value of the crystal frequency is exactly 11.0592 MHz.

You can also use 12 and/or 16 MHz crystals. If you do, modify the timer registers' contents to return to the conventional baud rates.

## The I²C bus output

When you use standard microcontrollers without integrated I²C hardware interface, you need to generate the protocol in assembly language (or high level language if supported). This is sometimes referred to as 'bit banging'.

We are free to choose which port will serve as the I²C lines as long as they are open-drain. Because of the internal architecture of the 8051, it makes sense to use port 1. Note that the **hardware** interface of PHILIPS SEMICONDUCTORS 80 C652 microcontroller (pin to pin compatible with 80 C52) brings out the I²C SCL clock on port 1.6 (pin 7) and the SDA data on port 1.7 (pin 8). Why not take advantage of future compatibility?

Two **pull up** resistors take the bus to the high state at rest. We have placed two other resistors as series protection against over voltages on the SCL and SDA lines (recommended especially to protect against ESD).

## Access to the Memory Field

All of the memory fields that we are using in our modules are external to the microcontroller. The microcontroller is capable of recognizing and controlling two types of memory space configuration:

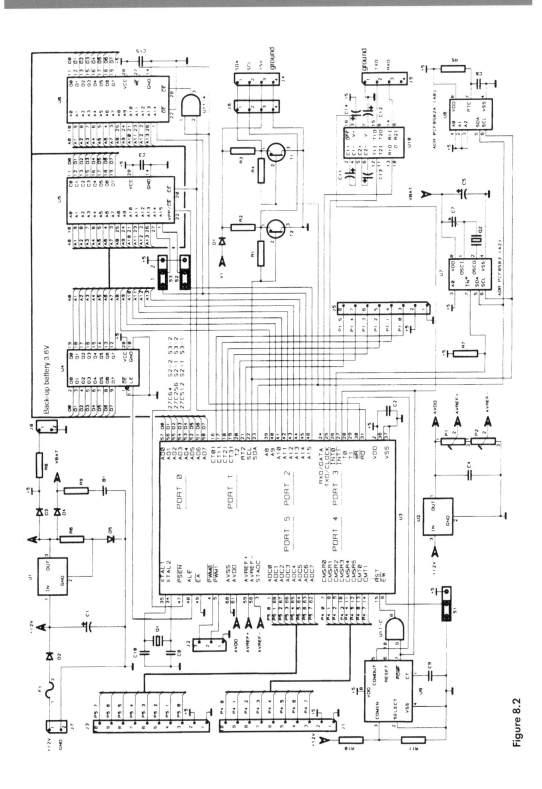

Figure 8.2

1. The RAM alone mode: only RAM present.

2. The 'RAM/ROM' mode: RAM's and (EP)ROM's simultaneously present.

We have chosen the second configuration. To implement the RAM/ROM mode, we need complementary circuits to access all of the memory space. The internal architecture of the 8051 microcontroller allows you to cover twice the memory space (64K + 64K) with the same number of addresses (64K), thanks to a structure of overlapping RAM and ROM address fields. First, you need to specify where you want to go in the memory field (either RAM or ROM), then specifically address the particular memory type.

## Demultiplexing and Selecting the Memory Zones

Decoding of the three highest order address lines of the microcontroller (A15, A14, A13) enables you to cut up the entire memory field (RAM and ROM) into 8 selectable fields of 8 Kbytes each.

You can demultiplex and decode the memory address field with a 74LS(HC)138 circuit, whose truth table and address zones are shown in Figure 8.3. Each of this circuit's outputs goes towards the memory ICs' CE or CS (chip enable or chip select), each consisting of 8 Kbytes of RAM and

| Address decoder (...138) | | | | | Memory fields (address values) | |
|---|---|---|---|---|---|---|
| Inputs | | | Outputs | | | |
| | | | Active outputs | Pins | From | To |
| H | H | H | Q7 | 7 | E000 | FFFF |
| H | H | L | Q6 | 9 | C000 | DFFF |
| H | L | H | Q5 | 10 | A000 | BFFF |
| H | L | L | Q4 | 11 | 8000 | 9FFF |
| L | H | H | Q3 | 12 | 6000 | 7FFF |
| L | H | L | Q2 | 13 | 4000 | 5FFF |
| L | L | H | Q1 | 14 | 2000 | 3FFF |
| L | L | L | Q0 | 15 | 0000 | 1FFF |

Figure 8.3

8 Kbytes of EPROM. CE and CS have identical addresses. You will need to specify one.

## Selecting appropriate memory types

To access the correct RAM or (EP)ROM memory type, the microcontroller issues address and control signals which perform memory selection. In addition, in order to reduce the number of pins on the IC, data and lower order address bits are time-multiplexed on the very same bus. Obviously, you need to perform the reverse operation to separate the address and data information, and to present them on the appropriate address and data bus. This is accomplished with the help of an LS or HC 573 circuit, whose purpose is to instantly latch the lower-order address bits on command from a service signal coming from the microcontroller (the 'ALE' signal, or 'address latch enable'). This 'trick' saves 8 pins of the microcontroller!

# 9

# The Standard I$^2$C Modules

## General Remarks

Now it's time to introduce the first I$^2$C modules and explore their functions in depth! All of the modules that we propose have a wide range of applications, and are illustrative only. Here are some applications of I$^2$C modules:

Analog to Digital conversion is useful for all kinds of things – You could even measure the temperature of your bedroom or your bathtub! In conjuction with other I$^2$C modules, you could then display the information on an LCD or on your TV, and even control the temperature over the I$^2$C bus via a remote control.

Whatever you choose as your application, in the end it will boil down to communicating with our modules, acquiring information and processing it, using the I$^2$C bus as our data highway.

We leave it up to you and your creative ability to invent your own application. We will show you how to take hold of the information, how to manage it (in the broadest sense of the term), and how to use the results to control the output.

## Personalizing the I$^2$C modules

The I$^2$C modules that will allow you to customize your system may be functionally independent of each other, but all of them can be interconnected on the same I$^2$C bus.

We have decided to use standard commercially available types of integrated circuits in the principal modules that we have chosen to use in these I²C systems. The table in Figure 9.1 gives the values of the addresses at which you can find them. Note the two following important details:

♦ The crosses correspond to **re-configurable** bits of the circuit I²C bus addresses. It's up to you to arrange them for your own convenience at the wiring level of the PCB itself. Note that we have presented many possibilities on each module; you can use several circuits of the same type so long as you give them different addresses.

♦ Take care that the value of the circuit's address does not change when the circuit is called by the software (not usually a problem when the address pins are hard-wired). The LSB of the first byte transmission of any I²C transaction can be either 1 or 0 depending on whether the transmission is a READ or a WRITE operation.

Bytes to be loaded in address register of μC
= address code (7 bits) + 1 bit of R/W

| FUNCTIONS | | REF | ADDRESS (7 bits) | + R/W (1 bit) | = byte (8 bits) |
|-----------|-----|-----|------------------|---------------|-----------------|
| Clock/event counter | | PCF 8573 | 1 1 0 1 0 x x | 1/0 | 1 1 0 1 0 x x 1/0 |
| | | PCF 8583 | 1 0 1 0 0 0 x | 1/0 | |
| Memories | RAM 265 x 8 | PCF 8570 | 1 0 1 0 x x x | 1/0 | |
| | RAM | PCF 8570C | 1 0 1 1 x x x | 1/0 | |
| | 128 x 8 | PCF 8571 | 1 0 1 0 x x x | 1/0 | |
| | E2PROM 128 x 8 | PCF 8582A | 1 0 1 0 x x x | 1/0 | |
| | Input/output | PCF 8574 | 0 1 0 0 x x x | 1/0 | |
| | | PCF 8574A | 0 1 1 1 x x x | 1/0 | |
| | A/D-D/A | PCF 8591 | 1 0 0 1 x x x | 1/0 | |
| Drivers | LED | SAA 1064 | 0 1 1 1 x x x | 1/0 | |
| | LCD | PCF 8577 | 0 1 1 1 0 1 0 | 0 | |

Figure 9.1

## Standard module address table

We suggest that you create your own table by replacing the crosses in ours with your own values. This will prevent finding two 'tenants' at the same address.

Figure 9.2 gives you the general table for address attribution for the principal components.

| | FUNCTIONS | REF | ADDRESS (7 bits) | + R/W (1 bit) | = BYTE TRANSMIS. (8 bits) |
|---|---|---|---|---|---|
| Memories | Clock/event counter | PCF 8583 | 1 0 1 0  0 0 0 | 1/0 | A.0 |
| | RAM | PCF 8570 | 1 0 1 0  0 0 1 | 1/0 | A.2 |
| | E2PROM | PCF 8582A | 1 0 1 0  0 1 0 | 1/0 | A.4 |
| | Input/output | PCF 8574 | 0 1 0 0  0 0 0 | 1/0 | 4.0 |
| | A/D-D/A | PCF 8591 | 1 0 0 1  0 0 0 | 1/0 | 9.0 |
| Drivers | LED | SAA 1064 | 0 1 1 1  0 0 0 | 1/0 | 7.0 |
| | LCD | PCF 8577 | 0 1 1 1  0 1 0 | 0 | 7.4 |

**Figure 9.2**

A number of standard sub-functions are common in all of these applications:

+ Digital input/output ('I/O'): monitoring of 'YES' or 'NO' information, such as whether or not a switch is closed or a tank overflows; or controlling a contact, turning on an LED, turning off a relay, starting or stopping a motor, or reading a digital number presented at the port (via a DIP switch, for example).

+ Analogue/digital conversion: measurement of the size of a physical quantity (temperature, pressure . . .), proportional control; transformation of physical amplitudes into numerical values for calculation.

+ Digital/analogue conversion: creation of particular control voltages to control dc motors or LCD contrast.

+ Real time clocks and event counters: counting the passage of time, chronometry, periodic alarms for safety applications, system energy conservation, time and date stamp for point of sales terminals or bank machines....

# Four Digit LED Display Module

Here's your first module which we chose because it is extremely easy to make and use, and has a very useful function. This module contains a simple LED driver and four digit LED display, and serves as a direct human interface – it indicates operational results via the LED's. It can also make it easier for you to debug future system errors by indicating them on the LEDs! A display is always useful, at some time or another.

Let's look at the circuit diagram of this module (see Figure 9.3).

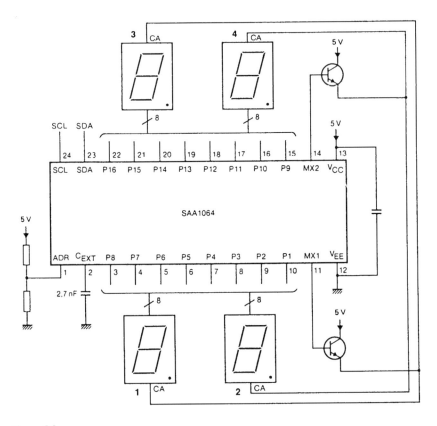

Figure 9.3

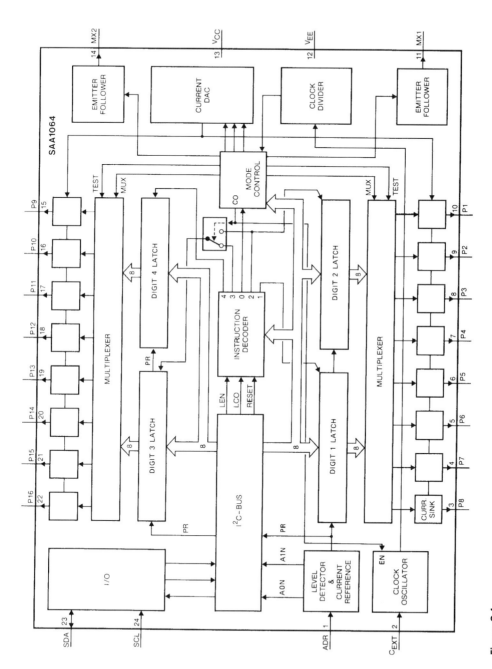

**Figure 9.4**

The entire system is concentrated in a single twenty-four pin integrated circuit: the SAA1064 of Philips Semiconductors, whose block diagram is shown in Figure 9.4.

Despite the relatively small number of connections, this integrated circuit is capable of controlling thirty-two segments (two pairs of two seven-segment digits and a decimal point) with common anodes in a duplexed display mode, using two PNP BC 547 transistors (or the equivalent), which can supply up to 21 mA. LED drive being a common function, there would be nothing very original about this IC... except that it is controlled by the I²C bus!

Let's look a little closer at this ICs characteristics. What do we have to transmit to make it work? This circuit's particular protocol are presented in Figure 9.5.

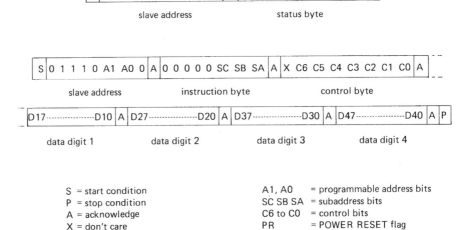

| S | 0 | 1 | 1 | 1 | 0 | A1 | A0 | 1 | A | PR | 0 | 0 | 0 | 0 | 0 | 0 | 0 | 1 | P |

slave address                          status byte

| S | 0 | 1 | 1 | 1 | 0 | A1 | A0 | 0 | A | 0 | 0 | 0 | 0 | 0 | SC | SB | SA | A | X | C6 | C5 | C4 | C3 | C2 | C1 | C0 | A |

slave address              instruction byte                control byte

| D17 --------------- D10 | A | D27 --------------- D20 | A | D37 --------------- D30 | A | D47 --------------- D40 | A | P |

data digit 1              data digit 2              data digit 3              data digit 4

S = start condition          A1, A0   = programmable address bits
P = stop condition           SC SB SA = subaddress bits
A = acknowledge              C6 to C0 = control bits
X = don't care               PR       = POWER RESET flag

**Figure 9.5**

Despite having only a single external address pin, you can address up to four modules of the same type in the address byte! Is this for real? How is this possible?

The ADR pin is controlled by a DC voltage that defines the value of the two address reconfiguration bits, A0 and A1. It accomplishes this thanks to a little internal A/D converter when it is compared to internal reference voltage thresholds. (This is an exceptional way to define an I²C address: usually the address pins are purely digital!). With this 'trick', the designers of this IC achieved the function of 2 pins in 1!

The table opposite shows the relationship between all of these elements.

You can power the ADR pin on your module to the level of your choice, thanks to the space left on the circuit board for a resistance divider

Table 9.1

| | Bits | | Address value | Address value |
|---|---|---|---|---|
| ADR voltage | A0 | A1 | (write) | (read) |
| $V_{ee}$ | 0 | 0 | 70h | 71h |
| 3/8 de $V_{cc}$ | 0 | 1 | 72h | 73h |
| 5/8 de $V_{cc}$ | 1 | 0 | 74h | 75h |
| $V_{cc}$ | 1 | 1 | 76h | 77h |

or straps. We have decided to connect it to VCC, naming it 76h for write cycles. Figure 9.6 gives the values of the data to be transmitted for segment display and the numbers on one of the SAA 1064 digits.

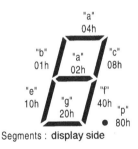

Segments : display side

Figure 9.6

## Instruction word

There is an instruction byte after the byte declaring the slave address. This byte's function is slightly more complex than that of the preceding byte. Its simple presentation hides a few tricks, which can be declared with the help of SA, SB, SC bits, which represent this word's low order bits, as indicated in the protocol.

These three bits determine the value of a **pointer**. The pointer's purpose is to indicate to the SAA1064 into which of its internal registers the data byte will be written (the data byte immediately follows the instruction byte). Once this task has been accomplished, all of the other bytes that follow will be gathered up and stored in registers whose sub-addresses follow, (according to the sequence shown in Table 9.2).

**Table 9.2**

| SC | SB | SA | Sub-address | Function |
|----|----|----|-------------|----------|
| 0 | 0 | 0 | 00 | command register |
| 0 | 0 | 1 | 01 | digit 1 |
| 0 | 1 | 0 | 02 | digit 2 |
| 0 | 1 | 1 | 03 | digit 3 |
| 1 | 0 | 0 | 04 | digit 4 |
| 1 | 0 | 1 | 05 | unused |
| 1 | 1 | 0 | 06 | unused |
| 1 | 1 | 1 | 07 | unused |

This particular process is called an **auto increment** of the sub-address and allows the master to rapidly initialize, modify and update information on the display with a single transmission. Auto increment accomplishes this by automatically incrementing the value of the pointer after each data byte transmission, saving time and increasing execution speed! To simplify our first example, we have chosen the value 00h for the first location.

## Control word

This 8-bit word is only defined for the seven lower order bits (C0 to C6). The most significant bit, C7, is not used (yet)! Each of these bits has its own life, independent of the others. Their actions are summarized in Table 9.3. and, in our module:

**Table 9.3**

| Display mode bit |
|---|
| $C0 = 0$    static display digits 1 and 2 |
| $C0 = 1$    dynamic display digits $1 + 3$ and $2 + 4$ |
| $C1 = 0/1$ digits $1 + 3$ on/off |
| $C2 = 0/1$ digits $2 + 4$ on/off |

| Test bits |
|---|
| $C3 = 1$   (TEST) = all segment |

| Brightness control bit |
|---|
| $C4 = 1$    current 3 mA ⎫ |
| $C5 = 1$    current 6 mA ⎬ 21 mA maximum |
| $C6 = 1$    current 12 mA ⎭ |

**Table 9.4**

| C7 | C6 | C5 | C4 | C3 | C2 | C1 | C0 | |
|---|---|---|---|---|---|---|---|---|
| – | 1 | 1 | 1 | 0 | 1 | 1 | 1 | =77h |

# Data word

In these bytes, you declare the digits that you want to display.

*Example:* To display a 9, you have to select the segments:
Table 9.5 shows the standard codes corresponding to each digit.

**Table 9.5**
**(a)**

| "d" | ..................................................... : | 02h |
|---|---|---|
| "b" | ..................................................... : | +01h |
| "a" | ..................................................... : | +04h |
| "c" | ..................................................... : | +08h |
| "f" | ..................................................... : | +40h |
| "g" | ..................................................... : | +20h |
| | ..................................................... : | 6Fh |

**(b)**

| to obtain: | | you must load via I²C bus |
|---|---|---|
| "1" | ..................................................... : | "48" |
| "2" | ..................................................... : | "3E" |
| "3" | ..................................................... : | "6E" |
| "4" | ..................................................... : | "4B" |
| "5" | ..................................................... : | "67" |
| "6" | ..................................................... : | "77" |
| "7" | ..................................................... : | "4C" |
| "8" | ..................................................... : | "7F" |
| "9" | ..................................................... : | "6F" |
| "0" | ..................................................... : | "7D" |
| decimal point | ..................................................... : | "80" |

# Summary

To brightly display '1 2 3 4' (21 mA), in multiplexed mode, you must send, the following series of bytes via the I²C protocol:

| Byte number | Signification | Values |
|---|---|---|
| byte no. 1 | address (pin to VCC) | 76h |
| byte no. 2 | instruction word | 00h |
| byte no. 3 | command word (multiplex, 21 mA) | 77h |
| byte no. 4 | data word to display "1" | 48h |
| byte no. 5 | data word to display "2" | 3Eh |
| byte no. 6 | data word to display "3" | 6Eh |
| byte no. 7 | data word to display "4" | 4Bh |

# Input/Output Module

Now, for the next module based on an interesting and versatile member of the I²C family, the PCF8574 'I/O Expandor'....

The PCF8574 is an 8-bit bi-directional latched port which can be Written to or Read via the I²C bus. Once we explain how to do this, you will then know how to do most things on the bus!

# Digital input/output – the PCF8574

There are 3 modules shown in the back of this book that use the PCF8574 in three different ways:

+ a stripped-down version that allows you to access simple digital ports;

+ a variation that allows you to use the module as a logical information display via LED's;

+ a third version onto which you can mount components that operate as interfaces between low power logic and low power control.

Let us look, in detail, at the operation of this module – there are only two components! It's no secret that the I²C bus reduces the number of components and pins.

### Operating principle of the PCF8574 I/O Expandor

The block diagram of the Philips Semiconductors' PCF8574 is shown in Figure 9.7, and its pinout in Figure 9.8.

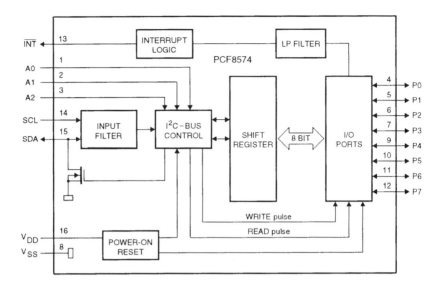

**Figure 9.7**

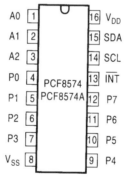

**Figure 9.8**

Although this component is a slave, it can either be written into or it can be read. To activate a bit on its output pins, this component is written to. It can be read to determine what is at its port. The PCF8574 also has a useful interrupt output pin which is activated when a value on the port changes. This can be very useful if you want to interface it to a control keyboard without having to constantly poll, as we will show later on.

### Extra

We have included two circuits on each module. You have the option of disconnecting one with switches.

### Circuit addresses

You can choose the 'names' of your circuits by specifying their addresses. There is a clever set of holes in the printed circuit layout that enable you to configure up to eight identical circuits (four modules of the same kind).

**Note:** It is entirely possible to have many more PCF8574's (hundreds even!) by dynamically time-multiplexing by logic the A2, A1, A0 addresses via other PCF8574 's (see Figure 9.9). This is just one of a number of tricks to increase the slave address range of an I²C device.

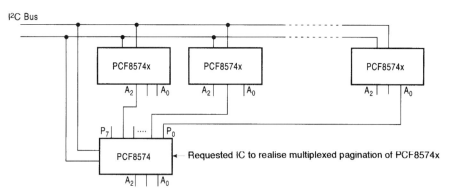

**Figure 9.9**

### Maximum output current

This quasi-bi-directional output port is latched. It maintains the information that it receives and, is capable of supplying in its low state a current that can directly drive LED's. The port can sink 25 mA; nominal LED current is usually around 10 mA.

### Reset

At power up, a reset device sets the output port to the high state (1111 1111 = FFh), enabling you to directly write into it. Writing all 1's to the PCF8574 is required to put it into the 'Read' mode.

# Writing/reading at the port

## Writing

The PCF8574's port bits are entirely independent and can be used either as input or output.

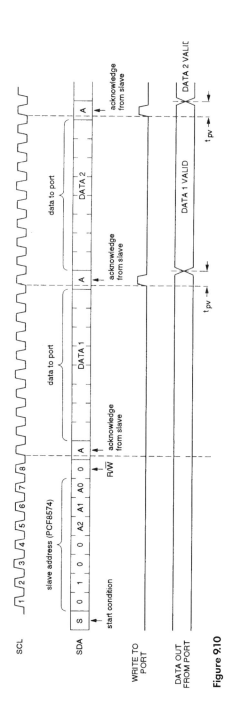

**Figure 9.10**

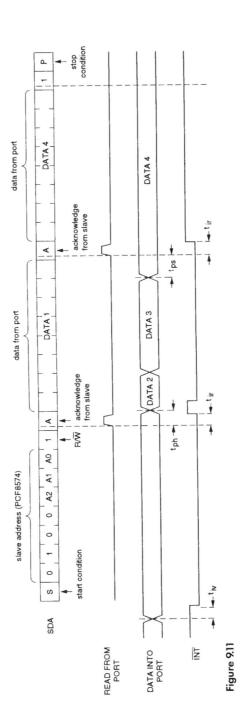

**Figure 9.11**

To write , the master (your microcontroller) first addresses the circuit by setting the last bit of the slave address to 0 (write mode); waits for the PCF8574 acknowledgment, then sends the data byte. The PCF8574 then presents the data it has just received on its port.

As shown in Figure 9.10, the data appear on the output port just after the acknowledgment of receipt indicating that the data have arrived. Writing a 1 in one of the data bits transfers a 1 to the corresponding bit of the output port after acknowledgment – for example, data bit 5 corresponds to output port bit 5. To set the entire 8-bit port to a state where it is waiting for data (input mode), it is necessary to transfer a data word of FFh.

The number of data words that you can send successively is not limited, in principle. Of course, each byte overwrites the previous one.

## Reading

Let's consider how to read the port. First, the port must be set to all 1's. Now we can transmit the slave address with LSB set to 1, indicating a read operation. The following data byte reflects the value on the port... or does it? Remember that information on the port can only be read so fast, and no faster! Even at 100 Kbit/sec I²C speed, the data on the port (now an input port) can change rapidly – even more rapidly than the bus can read it. For example, in Figure 9.11, the PCF8574 decides to really **read** what is on its port at the (last) moment, when it gets the **receiver** acknowledgment from the master. When the bus is operating at its maximum frequency of 100 kHz, the data presented to the input port must not change at a rate exceeding 10 kHz or information may be lost!

In many human interface applications, data varies slowly and you have all the time you need to scan the states of the various input ports by software.

After the last bit of the last data word is read at the end of the port read cycle, the master must send a **non-acknowledgment** (NACK) during the ninth SCL clock pulse so the PCF8574 will know that it should say *good-bye* and release the bus so that a STOP condition can be performed by the master.

## Interrupt

The circuit already knows how to tell you that its inputs have been modified via a separate interrupt output pin. When we decide to use this information, it's an entirely different story.

There are two ways to handle this interrupt with application software:

## Solution 1

You can pay no attention to this signal and instead, you check periodically that nothing has changed (the polling method). However, if a bit has changed twice at the same place since you last checked, you have a problem.

## Solution 2

You can take advantage of this signal. When the port is in read mode, as soon as one of the inputs detects a rising or descending edge, an interrupt signal INT is created. This depends on at least **one** pull up resistor to $V_{DD}$ for all of the interrupt pins (in the case of multiple PCF8574's), because the outputs are open drain and can be connected to form a wired OR over the entire set of circuits or modules. This INTerrupt signal allows the main program to be diverted to check what is happening on the ports without having to poll.

Since several circuits might have sent an INT simultaneously, all the circuits must polled and read. Only when the changed value on the right port has been read does **its contribution to the INT disappear** (Figure 9.12).

Since all of the INT pins are connected in **wired OR**, the interrupts are examined, one by one, until all have been serviced. Once the interrupts having been cleared, we then go back and pick up the program where we left off.

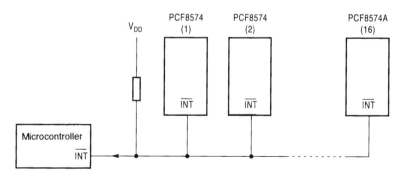

**Figure 9.12**

# Memory Modules (RAM and EEPROM)

There are different kinds of memories in the line of I²C bus compatible components:

♦ RAM

♦ EEPROM

♦ video memories

♦ FLASH memories

We will discuss the common small serial memories (RAM and EEPROM), which you will often use in your applications. EEPROM is particularly useful in applicatons where data retention during power-off is essential (for example: meter readings, electronic key, product identification number, etc).

A single pinning can be used for these ICs because they are very similar and their pinouts have been intentionally designed for interchangeability. As you can see, they are tiny DIL 8 packages (or SO ('Small Outline') for surface mounting. Regardless of the size of the memory, they are all accessed by the 2-wire I²C bus (Figure 9.13).

We will look at the PCF8570 RAM, organized in 256 × 8 bits. For EEPROM, we have chosen the PCF8582(x) family, organized in the same way. Note that there are many other series of I²C EEPROM's on the market (produced by XICOR, Excel, SGS-Thomson, and several others), which are not always compatible in every respect, but which offer a wide array of memory sizes and features to satisfy many applications.

## Memory addresses

As with the PCF8574, addressing is also re-configurable. The memories' respective addresses are:

| RAM | PCF | 8570 | : | 1 0 1 0 | A2 A1 A0 (R/W) |
|-----|-----|------|---|---------|----------------|
|     |     | 8570C | : | 1 0 1 1 | A2 A1 A2 (R/W) |
| EEPROM | PCF | 8582 | : | | |

## RAM

The PCF8570 256 × 8 bit RAM's specific protocol is shown in Figure 9.14.

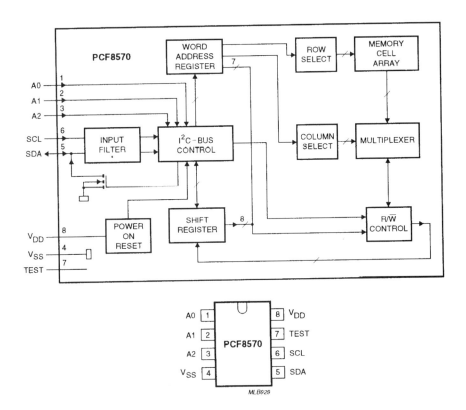

Figure 9.13

## Write mode

We address the PCF8570 by the usual Start Condition and Slave Address. After the component acknowledges us, we indicate the address at which the first data word is to be written. This step is essential! Thanks to an auto-increment feature (similar to the SAA1064 LED driver described earlier) which automatically increments to the next address, it is possible to input unlimited data which is stored in successively increasing RAM locations. The exchange is terminated by a Stop condition. When the highest possible address is arrived at, the circuit loops back and starts again from address zero.

## Read Mode

We have two variations of a read operation:

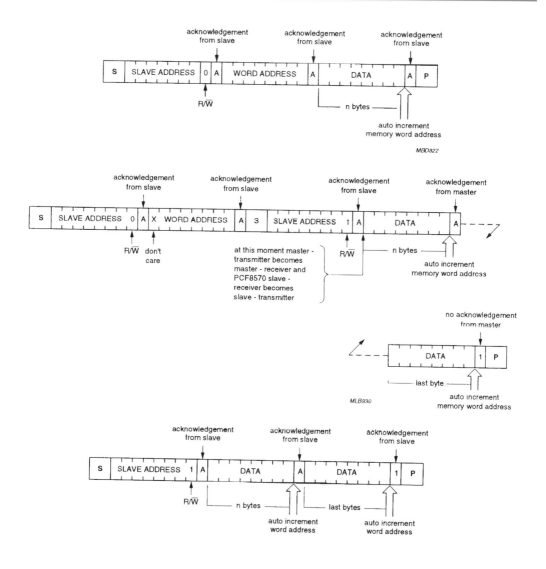

Figure 9.14

## Variant 1

The master reads the memory after presetting the read start address. The component is first addressed in write mode in order to indicate to the device from which address we want to begin reading. Without stopping (no STOP condition), a RESTART is generated (a second START condition), followed by the address of this same component, but this time in read mode. The component now knows where reading should begin because the address has been loaded into the auto-increment register. From this point on, the slave (the memory) sends data to the microcontroller, incrementing addresses until the master decides that it doesn't want any more. The microcontroller then sends a NON-acknowledgment (NACK software procedure) to inform the memory that it doesn't want to receive any more data. The transmission is terminated with a STOP condition, as usual.

### Variant 2

The master reads the memory directly after sending the slave address + read command. The component is addressed in read mode and sends data starting from the last incremented address contained in the auto-increment register, (which has been set by a previous write or read). This solution saves time in the exchange because you don't have to indicate the start address.

Note: At **power on reset**, the auto-increment register is initialized, so you can begin writing or reading at address zero. In general, however, make sure that an I²C component has an internal power on reset built in before assuming that the address pointer has a default value!

## EEPROM

The PCF8582(x), also known as Electrically Erasable PROM (EEPROM), allows you to store data (2 Kbits organized in 256 × 8), including your set points, temperature, alarms, and more, for a guaranteed minimum storage time of ten years in the absence of power. PCF8582(x) still allows you to change the values 100 000 to 1 000 000 times, while consuming only 10 micro amperes of current! (Many higher density types, for example Philips PCF 8594/98 4K and 8K EEPROMs have appeared, and some manu-facturers now make single chip I²C EEPROMs up to 32 Kbits!).

Its pinout is compatible with the preceding RAM (in case you want to replace one with the other). The block diagram of this circuit is shown in Figure 9.15.

### Write mode (see Figure 9.16)

The write procedure is identical to the procedure used for writing the PCF8570 RAM with one important exception. You can only send the circuit a **limited number of bytes**, instead of an unlimited number. The **erase/write** procedure (which takes about 30 ms per byte or 60 ms for both the erase/write), can only begin after the circuit receives and acknowledges

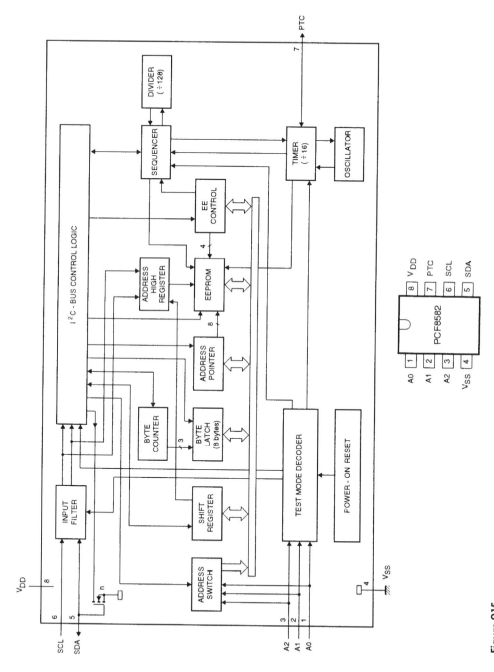

**Figure 9.15**

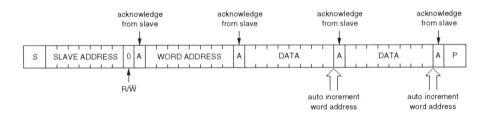

**Figure 9.16**

these bytes of data. The erase/write procedure is set in motion and controlled by an internal oscillator and timer.

The memory is inaccessible while it is programming the EEPROM cells. Even if you address it to reload bytes, it will refuse to send the acknowledgment that you are waiting for until programming is finished.

Here is a fundamental notion of the I²C bus – in addition to the standard acknowledgment of **proper reception**, we also have an acknowledgment in **real time** only after the function performed by the circuit is completed.

## Read mode

Read mode presents many of the same issues discussed in the section on the RAM.

### Solution 1

The master reads the memory after presetting the read start address (see Figure 9.17a).

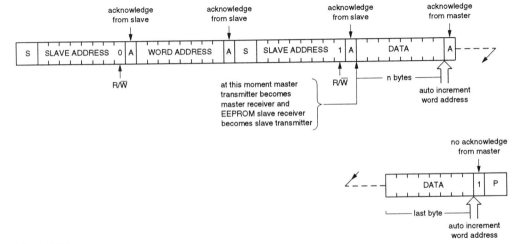

**Figure 9.17 (a)**

## Solution 2

The master reads the memory directly after the read command.

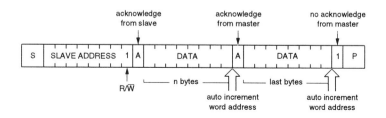

**Figure 9.17 (b)**

# A/D and D/A Conversion Module

Our analogue to digital (A/D) converters and digital to analogue (D/A) converter module uses the PCF8591, which is capable of converting **four** different analogue voltages to digital values for processing in the microcontroller. It can also generate **one** analogue voltage by converting an 8-bit digital value provided by the microcontroller.

This is not an arbitrary choice. You will frequently have to record several kinds of analogue information in your applications, such as temperature, pressure, battery level, signal strength, etc. These are digitally processed and can be subsequently displayed, used to control contacts, switches, relay etc. for example using the previously discussed I/O expandor PCF8574. The D/A output is useful for such jobs as LCD contrast control.

## The PCF8591 A/D-D/A converter

This practical circuit contains a 4-channel multiplexed A/D converter and one D/A converter, both 8 bits. You can use it to interface to many types of transducers. Like many other circuits in the same family, it has three address reconfiguration pins, enabling you to use eight of these circuits in the same system.

We have designed this module to accept two circuits, enabling you to customize their addresses, using **1001 A2 A1 A0 R/$\overline{\text{W}}$**, where 'Ax' are user defined.

## A/D conversion

There are four input pins that can receive analogue voltages. These are time multiplexed to a sample and hold circuit, and then converted by 8-bit successive approximation (see Figure 9.18).

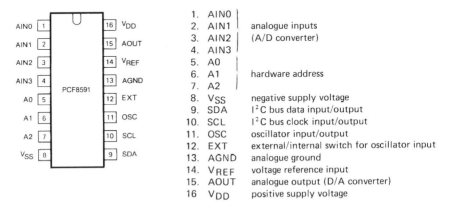

| | |
|---|---|
| 1. AIN0 ⎫ | |
| 2. AIN1 ⎪ | analogue inputs |
| 3. AIN2 ⎪ | (A/D converter) |
| 4. AIN3 ⎭ | |
| 5. A0 ⎫ | |
| 6. A1 ⎬ | hardware address |
| 7. A2 ⎭ | |
| 8. V$_{SS}$ | negative supply voltage |
| 9. SDA | I²C bus data input/output |
| 10. SCL | I²C bus clock input/output |
| 11. OSC | oscillator input/output |
| 12. EXT | external/internal switch for oscillator input |
| 13. AGND | analogue ground |
| 14. V$_{REF}$ | voltage reference input |
| 15. AOUT | analogue output (D/A converter) |
| 16 V$_{DD}$ | positive supply voltage |

**Figure 9.18**

Several reasons lead to this choice:

+ In many applications, especially those requiring human interface, there is plenty of time to convert the input analogue values.

+ The successive approximation technique is a bit slower than **flash** conversion (with its host of comparators), but it is considerably less expensive.

+ A faster A/D converter is not necessary as the maximum conversion rate is determined by the maximum speed of the I²C bus (100 Kbits/sec translates roughly to about 9K conversions per second – fast enough for telephone-quality speech!).

To save time during communication transfers on the I²C bus, you can use the auto-increment function during input multiplexing, enabling you to switch rapidly from one input to the next . You can also configure the inputs in single ended mode or take them two by two in differential mode, (which is practical for canceling components, temperature, and ageing drifts). The block diagram, Figure 9.19, shows the conventional SAR structure.

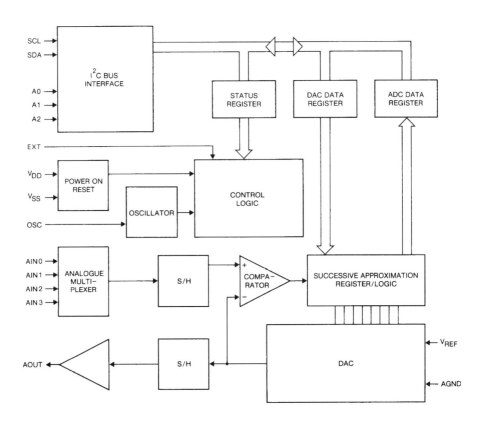

**Figure 9.19**

Although this may seem surprizing, you need a D/A converter for successive approximation A/D conversion. If we disconnect the return to the negative voltage comparator input, we are left with a D/A converter, which allows us to claim '*n*' A/D converters and '1' D/A converter for free! This is where the PCF8591 gets it's one D/A output.

# Circuit control via the I²C bus

## A/D conversion

The basic exchange protocol for an A/D conversion is given in Figure 9.20.

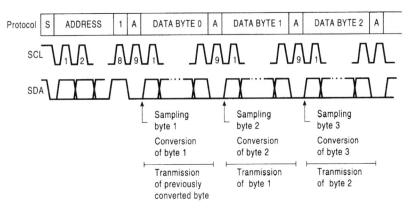

**Figure 9.20**

## Address byte

The address byte (Figure 9.21) can be re configured with the three A0, A1 and A2 pins. The R/$\overline{W}$ exchange direction bit must have the value of 1 for an A/D conversion and 0 for a D/A conversion.

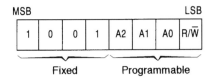

**Figure 9.21**

The second (**control**) byte requires some attention.

## Control byte

The control byte (Figure 9.22) is stored in an internal circuit register on arrival, and controls the various functions which can be performed by the PCF8591.

| MSB | | | | | | | LSB |
|---|---|---|---|---|---|---|---|
| 0 | X | X | X | 0 | X | X | X |

**Figure 9.22**

The four highest order bits manage the analogue inputs and outputs' operating mode(s); the four lowest order bits control the supply of the analogue inputs to be converted.

It is easiest to start with bits 4 and 5. These bits control the configuration of the inputs, as Figure 9.23 clearly shows. Setting bit 6 of this byte to 1, allows you to activate the D/A conversion output. The 3 lowest order bits are organized for other purposes. The 0 and 1 bits select the input channel. Bit 2 activates/deactivates autoincrementing.

To save you precious CPU processing time you can use auto-increment of the channel value after each A/D conversion to read each input in turn (by setting bit 2 to '1'). The 3 and 7 bits have no user function and should

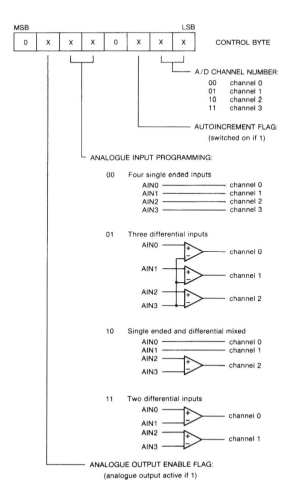

**Figure 9.23**

be set to 0. Be careful to avoid illegal states, for example the conversion of analogue channel number three (by setting bits 0 and 1 to '1'), if it has not been foreseen in the input configuration (for example, with bits 4 an 5 set to '1' and '0').

With respect to timing, Figure 9.24 shows the coincidences between the input signals and the resulting bytes in the course of A/D conversion. Again, the conversion time of this circuit depends only on the maximum speed of the I²C bus and not on the conversion principle chosen.

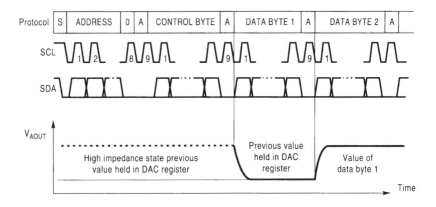

**Figure 9.24**

Figures 9.25(a) and (b) show the values of the bytes obtained, depending on the modes that you have chosen for the converter inputs (single ended or differential).

Note that , the codes in the initial byte are in two's complement mode in this case.

## D/A conversion

The protocol of this exchange via the I²C bus is shown in Figure 9.26; the value of the resultant dc voltage is given in Figure 9.27.

The intrinsic quality of the result depends (for A/D conversion) on the quality of the reference voltage's value.

In this type of circuit, everything often depends on the care you give to this reference voltage's generation (its value, its stability as a function of temperature). It is from this voltage, via resistance dividers, that the elementary parts which make up the output voltage are created.

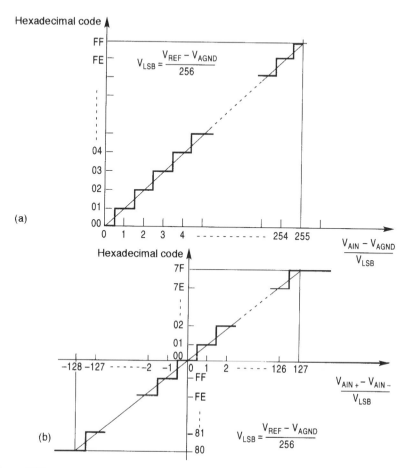

Figure 9.25

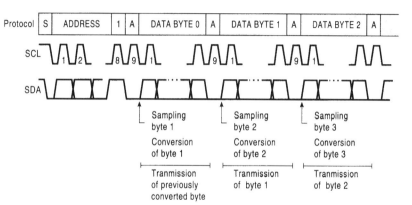

Figure 9.26

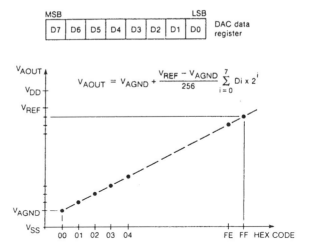

**Figure 9.27**

Figure 9.28 shows an application you can choose to regulate a heating system by measuring the temperature with a resistance bridge for better stability.

# Real Time Clock Module

Many applications require a **real** time clock for energy regulation, point of sales time stamp, security systems, elapsed time measurement, or simply for periodic system wake-up. There are three integrated circuits compatible with the I²C bus that satisfy the combined function of **clock and/or event counter**: the PCF8573, PCF8583, and PCF8593. The PCF8573 is a basic clock/calendar, the PCF8583 includes scratchpad RAM, and the PCF8593 is optimized for low power consumption.

We will describe the PCF8583. This circuit is organized much like the PCF8570 RAM. All time keeping registers, alarms, control and scratchpad registers are arranged as an 256 × 8 byte array.

# The PCF8583 RTC IC

This RTC circuit has low power consumption so that it can be easily backed up by a small back-up battery, or even capacitor. In addition, the clock function and memory retention are guaranteed for voltages as low as 1.0 V (at only 2 microamps!).

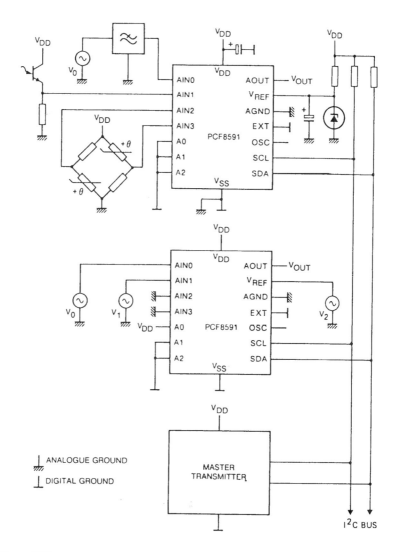

**Figure 9.28**

The circuit operates as a real time clock/calendar; in 12 or 24-hour mode. A 2-bit counter is also included to keep track of leap year. It also has an interrupt output, which is very useful in applications where the host microcontroller is 'put to sleep' until a pre-designated time. 240 bytes of scratchpad RAM is also available for storing of any application information (for example, telephone numbers, meter readings, wakeup times, etc). Its block diagram and pinout are shown in Figure 9.29.

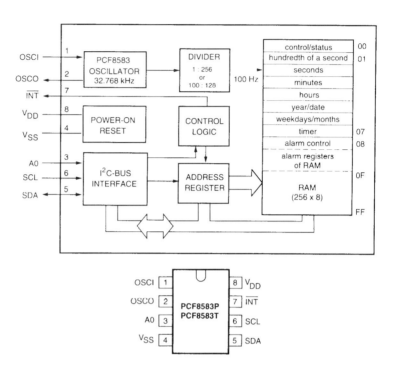

**Figure 9.29**

The application circuit diagram that we have chosen for our module is given in Figure 9.30.

As with many clock ICs, you must use 32.768 Hz quartz crystal frequency as the timebase.

Some times the fewer the number of pins on an integrated circuit, the more complicated it is to use, especially if it is a high performance circuit with loads of options like the PCF8583!

You have to be very careful, therefore, when you chat with this circuit via the I²C bus. Its I²C address is given in Figure 9.31, which illustrates the possibility of two sub-addresses by the A0 pin. We have retained this option to enable us to have two integrated circuits on the module, one for the clock function and the other as an event counter, for example.

The transmission protocol given in Figure 9.32 is identical to the PCF8570 RAM and PCF8582 EEPROM memories' protocol. Before talking to the PCF8583 RTC, you should know exactly what you want the circuit to do.

As shown in the PCF8583's block diagram, the first words in the memory zone are organized into dual-function registers, depending on whether the IC is in **clock or event counter** operating mode (Figure 9.33).

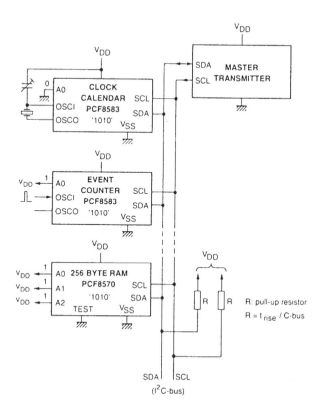

**Figure 9.30**

| 1 | 0 | 1 | 0 | 0 | 0 | A0 | R/W̄ |
|---|---|---|---|---|---|----|-----|

| Group 1 | Group 2 |

**Figure 9.31**

These registers occupy addresses from 00h to 0Fh. If you only use the clock mode, you can ignore the event counter registers (and vice-versa).

In addition to the **conventional** registers for this type of application (minutes..., alarm...), there is one that is truly special – the register located at the zero address, whose mission is the overall control of all of the PCF8583's operating modes. This is called the 'Control/status register'.

This register provides a number of application possibilities. We begin simply by:

* ignoring the 6 and 7 bits by setting them to zero (used only in event counter mode)

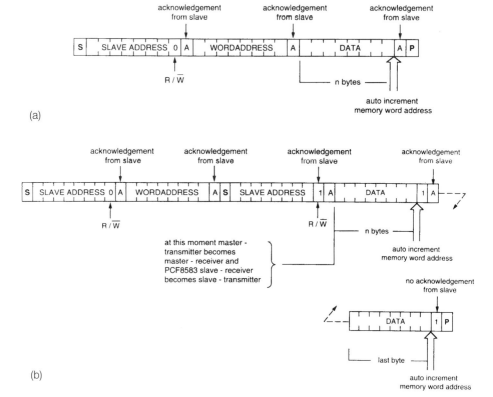

(a)

(b)

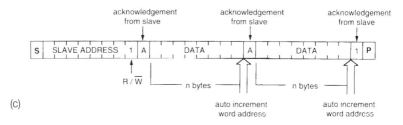

(c)

**Figure 9.32**

• deciding that we want to use the clock function (bits 4 and 5 set to zero

• setting bit 2 to zero to deactivate the alarm (at least to start with).

The word to be loaded at address zero will then be:

• either 00h to start (accomplished automatically at power up), to start the circuit in a clock mode, no alarm

• or (then) 04h to authorize an alarm.

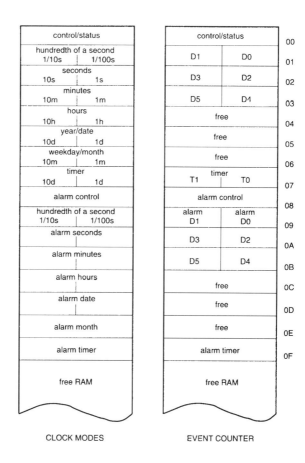

| control/status | | | control/status | | 00 |
|---|---|---|---|---|---|
| hundredth of a second 1/10s &#124; 1/100s | | | D1 &#124; D0 | | 01 |
| seconds 10s &#124; 1s | | | D3 &#124; D2 | | 02 |
| minutes 10m &#124; 1m | | | D5 &#124; D4 | | 03 |
| hours 10h &#124; 1h | | | free | | 04 |
| year/date 10d &#124; 1d | | | free | | 05 |
| weekday/month 10m &#124; 1m | | | free | | 06 |
| timer 10d &#124; 1d | | | timer T1 &#124; T0 | | 07 |
| alarm control | | | alarm control | | 08 |
| hundredth of a second 1/10s &#124; 1/100s | | | alarm D1 &#124; alarm D0 | | 09 |
| alarm seconds | | | D3 &#124; D2 | | 0A |
| alarm minutes | | | D5 &#124; D4 | | 0B |
| alarm hours | | | free | | 0C |
| alarm date | | | free | | 0D |
| alarm month | | | free | | 0E |
| alarm timer | | | alarm timer | | 0F |
| free RAM | | | free RAM | | |
| CLOCK MODES | | | EVENT COUNTER | | |

**Figure 9.33**

You have powered the circuit and loaded the 00h contents at the 00h address. What is remarkable is that it is already doing lots of things that you don't know about yet! Using a read routine, the hundredths of a second, the seconds and the minutes can be checked at the 01h, 02h or 03h addresses via the bus, on the other hand, the hour register located at 04h, is much more mischievous.

The hours register (Figure 9.34) is initially set to 00h at reset, which means that clock operates in 24-hour mode (bit 7 is 0). If we set bit 7 to 1, then 12 hour mode is activated and bit 6 then indicates AM or PM.

The tens hour digit is accessible on bits 5 and 4 and the units digit on the four 0, 1, 2, and 3 lower order bits, coded in BCD. The procedure is similar for registers 04, 05 and 06 as shown in Figure 9.35.

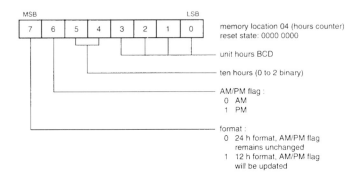

**Figure 9.34**

# Alarms

We now arrive at the critical stage of alarms. Register 08h, the 'Alarm control register' is the orchestra conductor (Figure 9.36).

The Alarm control register only allows you to choose the mode by which you trigger something. With the help of this register, you can preset your choice without setting off the alarm itself.

An alarm needs three elements to work:

♦ to know precisely the exact moment at which it should trigger

♦ to be activated or deactivated

♦ to put out control information to the external world when the alarm occurs.

### When should it work?

First load your alarm time into registers 0Ah to 0Fh. You may have noticed that there is a constant **offset** of 08h between the addresses where the instantaneous value of a variable (hour, minute,...) are stored and its associated alarm's values.

**Example:** hour address (04h) = 08h) = alarm hour address (0Ch).

### To be activated

After selecting the alarm conditions (register 08h) and loading your values (registers 0Ah to 0Fh), you can trigger an alarm via the Control/status

| Address register | Content | Meaning |
|---|---|---|
| 00H | 00H | I am a clock<br>with a disable alarm |
| 01H | 00H | 0 1/100s |
| 02H | 00H | 0 s |
| 03H | 43H | 43 minutes |
| 04H | C2H | |
| | :: | hour unit: 2 |
| | :: | because it's 2 o'clock (pm) |
| | : | |
| | : | |
| | : | bits 7 6 5 4 |
| | | 1 1 0 0 |
| | : | : : : : |
| | : | |
| | : | |
| | : | PM |
| | : | |
| | : | format<br>over 12<br>hours |
| 05H | 61H | |
| | :: | |
| | :: | day unit: 1 |
| | : | |
| | : | bits 7 6 5 4 |
| | | 0 1 1 0 |
| | : | : : : : |
| | : | |
| | : | year 89 is the<br>first after |
| 06H | D0H | |
| | :: | |
| | :: | month unit: 0 |
| | : | |
| | : | bits 7 6 5 4 |
| | | 1 1 0 1 |
| | : | : : : : |
| | : | |
| | : | day of the<br>week<br>Saturday: 6 |

**Figure 9.35**

register's bit 2 (remember this register is located at address 00h). Now, bit by bit comparisons take place in all of the clock registers concerned, until the 'big event' occurs . . . .

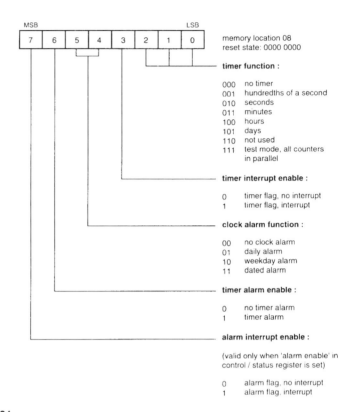

**Figure 9.36**

## Alarm: interrupt generation

The INTerrupt pin has been provided to interrupt the processor when an alarm occurs (if enabled via bit 7 of the Alarm control register). Since this output is an open drain, don't forget to install a pull-up resistance to V$_{DD}$. Before you activate the interrupt, you need to know how to terminate it, or your interrupt will never go away and your system may be permanently blocked!

## Activation

The master interrupt enable switch is 2 bit of the control/status register. We can choose either to have an alarm condition simply set the 'alarm flag' (bit 1), or in addition, trigger the external interrupt. If we choose to trigger the interrupt by setting bit 7 in the Alarm control register to 1, the INTerrupt output passes to the low state at the moment the alarm occurs (and remains there if you don't do anything about it)! At the same time, the alarm flag (bit 7 of the register located at address 08h) is set to the 1 state by the internal electronics.

## Deactivation

You need to reset the control status register's bit alarm flag to terminate the interrupt output. Of course you may want to first set up the conditions for the next alarm before doing so. Table 9.6 shows an example of how to initialize the PCF8583 RTC.

**Table 9.6**

| Address of the register | Contents | Meaning |
|---|---|---|
| 00h | 00h | Enter unit value, clock |
| from 00h to 07h | | |
| 01h | xxh | 1/100 s |
| 02h | xxh | s |
| 03h | xxh | minutes |
| 04h | xxh | hours, AM/PM... |
| 05h | xxh | date, year |
| 06h | xxh | days, months |
| then load alarms | | |
| 09h | bla-bla h | |
| 0Ah | bla-bla h | |
| at | | |
| 0Fh | bla-bla h | |
| now wait for alarm | | |
| 08h | 00h | |
| 00h | 04h | |
| wait for coincidence!!! | | |

**Note:** the manufacturer's component datasheet offers many other application possibilities, including a **timer and event counter** function which we have not discussed here. To understand the circuit's internal operation, see Figure 9.37.

Now that you know how to terminate the interrupt, we will use it to set our clock precisely.

When the circuit is first powered, all of the registers are initialized and the circuit starts up **naturally** at 0.00.00:00 on January 1.

All by itself, the circuit generates a 1 Hz signal on the interrupt pin for calibration. You can display its signal on an oscilloscope. To adjust the crystal oscillator frequency with precision, replace the fixed capacitor of 10

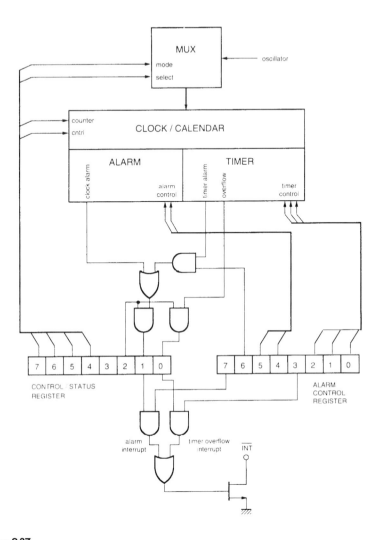

**Figure 9.37**

to 20 pF by a trimmer capacitor of 40 pF, (initially adjusted to 15 to 20 pF). Unfortunately 1 Hz is rather hard to see.

If you are in a hurry, you can write a brief sub-routine to activate the interrupt at regular intervals. You achieve this with another type of circuit operation, the timer mode, and by employing the following types of operations:

1. set the clock to time $T$,

2. set the alarm to time $T + dT$,

3. repeat the routine at each $dT$ (interrupt).

There is an alternative:

Start from the principle that the resonant element is piezo-electric and that the oscillator can be capacitively perturbed by an oscilloscope probe. Take the microphone from your stereo, then connect it to a good audio amplifier (with at least 35 kHz response). Place your microphone near the crystal; you will obtain a signal by acoustic coupling that you can easily connect to the input of a frequency meter!

# Keyboard and Character Stream Display Module

We propose a system for alphanumeric display keyboard control. We suggest multiple variations on the same theme, from which you can choose your preferred option.

## General

Let us begin, therefore, by examining the block diagram of the module shown in Figure 9.38. First, you will notice the component uniformity: the PCF 8574's belonging to the I²C family that we have already used and whose protocol has been presented in detail earlier in this chapter.

Until now, everything is just fine and all you need do is baptize them with the sweet little nicknames that you will choose, with the help of the usual sub-address bits (the well known pins A0, A1, A2). To be as consistent as possible in our explanations, we have chosen to call them, respectively, as follows:

IC1:    0 1 0 0 1 1 1 1 or 4Fh:    keyboard 1

IC2:    0 1 0 0 1 1 0 1 or 4Dh:    keyboard 2

IC3:    0 1 0 0 1 0 1 1 or 4Bh:    display 1

IC4:    0 1 0 0 1 0 0 1 or 49h:    display 2

That seems like quite a lot – four circuits to control a miserable twelve key keyboard and a liquid crystal display. Some would even go so far as to say that, with a little microcontroller, one could do that very well and we could not honestly say that they are entirely wrong..., but let's look in detail at what we are offering.

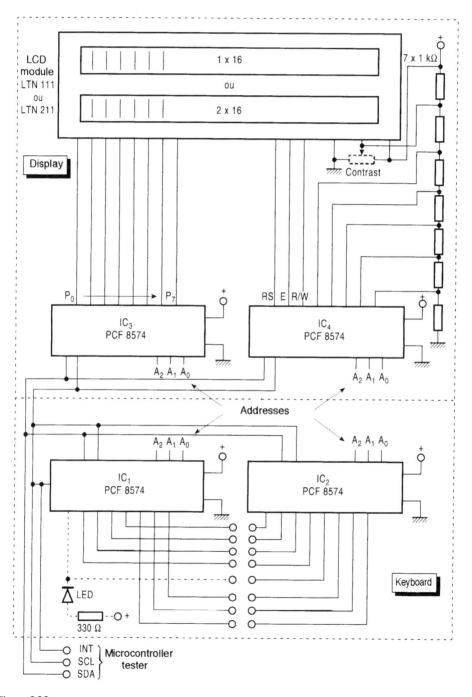

**Figure 9.38**

# Different hardware options

## First option: a simple one

It can be made using only the two circuits $IC_1$ and $IC_3$, twelve key telephone type keyboard (four rows and three columns) and a simple display. We have not been totally honest here because the display is not at all simple since, frankly, it is not a display but a high level display module that we have chosen.

Be assured that these models are used for many mass market applications and their price is entirely reasonable. Moreover. since they have some intelligence. their control is not very complicated. But we will get back to that later. This display – LTN 111xxx – allows you to display a row of sixteen characters (a thirty five point matrix for each one).

## Second option: a standard solution

Here we use three integrated circuits – $IC_1$, $IC_3$, $IC_4$, thanks to which you will be able (in addition to the function indicated) to modify the contrast of the display via the $I^2C$ bus, making use of very simple control software. It is the option that we have chosen to describe because of its favourable function/price ratio.

A sub-variant, hardly more costly and easy to use, is one that employs the LTN 211, which allows you to display two rows of sixteen characters and can be very practical for certain applications.

## Third option: de luxe!

You can always add a fourth circuit ($IC_2$) to the last version presented, which will allow you to use a 64 key keyboard ($8 \times 8$) for other than household applications, if you like.

**Important remark:**
The ensemble is connected to the CPU only by the $I^2C$ bus.

It can be physically mounted on the mother board of the CPU itself (mechanical format of the module identical in width to the others; it can be split down the middle so that you can separate the keyboard part from the display part).

Nothing prevents you from moving it a few metres, using a "wired" connection because the I²C bus has very few wires.

Later, you will be able to cut its umbilical cord and connect it to the CPU via other media (IR . . . ).

# Software options

## For the keyboard unit

We have brought out the interrupt pin from the PCF 8574, in case you want to operate your microcontroller in this mode. Remember that you will then need a pull-up resistance to the +5 V to use the signal and that you will also have to poll all of the circuits which might have generated interrupts, before going on with your program. In many applications, as we have already often pointed out, polling is time consuming and your microcontroller may spend a good part of its life asleep!

## For the display unit

The module has fourteen connection pins (Figure 9.39). Eight of them are reserved for data (D0 . . . to . . . D7), three for the control modes and the last three for positive and negative power supplies and the contrast control voltage.

Although the data are processed in eight bit groups, for reasons of compatibility with certain (old?) microcontrollers on the market, it is also possible groups of four bits each. In this case, you use only one of the two circuits $IC_3$ and $IC_4$, but this requires some unpleasant acrobatic manoeuvres at the software level. We have chosen another way out, which consists of working conventionally on groups of eight data bits (dedicating a PCF 8574 for that purpose) and handling the three remaining bits by another PCF 8574.

Five unused bits remain. Certainly not! Five usable bits – yes! We have used these bits, via the interposed I²C bus. for a remote contrast control.

OK, now that you have all of the options in mind, take a few minutes to think about your choice and then come back to us to find out how our display module operates.

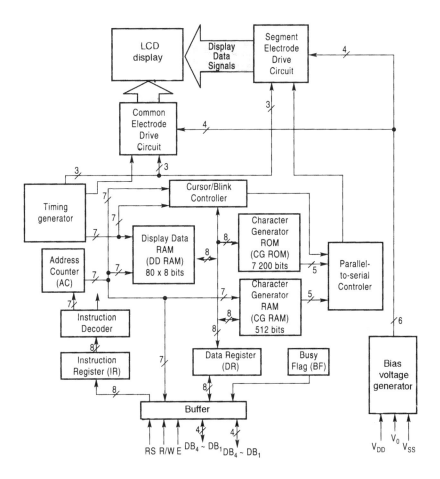

**Figure 9.39**

## The keyboard unit

### Hardware

Let's get back to our choice of using a single integrated circuit PCF 8574 for the coding/decoding function of a twelve key (3 × 4 matrix) keyboard. This circuit has a distinctive feature that we have hardly considered until now. Its input/output ports do not at all behave in the same way when they are in the high state as when they are in the low state. If one of the pins of the port is in the high 1 state, it can easily be forced to the low state, since it has the same electronic structure as a circuit output stage. In addition, when the circuit is waiting (in other words, when it is in input mode) it accepts any modification of the outside world as information. If it is a 1 that arrives, nothing changes, but if it is a 0, the port goes to zero for all of the time that the information is present on the pin in question.

*Nec plus ultra.* At this very moment, it lets you know by simultaneously generating a signal, and placing it on its interrupt pin, allowing you (if you so desire) to come a-running and grab hold of the something in question!

## Software

The organization of the keyboard in the form of three columns and four rows allows a single 8 bit port to read the twelve keys (we have kept aside the last bit in case we want to control an LED).

We have attributed all these goodies to the PCF 8574 as follows:

|      | P0 | P1 | P2 | ports |           |
|------|----|----|----|-------|-----------|
| Keys | 1  | 2  | 3  | P3    |           |
|      | 4  | 5  | 6  | P4    |           |
|      | 7  | 8  | 9  | P5    |           |
|      | *  | 0  | #  | P6    | and LED: P7 |

and the data byte that we will transmit to the integrated circuit via the I²C bus for initialization purposes at the beginning, will be:

| LED | LINES |    |    |    | COLUMNS |    |    |     |
|-----|-------|----|----|----|---------|----|----|-----|
| P7  | P6    | P5 | P4 | P3 | P2      | P1 | P0 |     |
| 1   | 0     | 0  | 0  | 0  | 1       | 1  | 1  | 87H |

which will represent the keyboard wait state, LED out (since it is wired to plus 5 V and the port P7 is in its high state). Now, if you depress the key 8, for example. P1 goes to 0 for all of the time that the key is depressed and also the interrupt immediately goes to 0.

If we stop there, we are no better off than we were before, because P1 has just gone to zero, yet if we had pressed the keys 2, 5 or 0, the same thing would have happened! We would have. in any case, obtained:

| LED | LINES | | | COLUMNS | | | | |
|-----|-------|-----|-----|---------|-----|-----|-----|-----|
| P7 | P6 | P5 | P4 | P3 | P2 | P1 | P0 | |
| 1 | 0 | 0 | 0 | 0 | 1 | 0 | 1 | byte no. 1 |

values that we will, nevertheless, read and transfer to the microcontroller via the I²C bus, where they will be carefully preserved in a memory location. (Duration of all of these operations is some hundreds of microseconds, because it is necessary to transmit four to five words via the I²C bus). It is, then, necessary to remove all doubt and uncertainty between the 8, the 5, the 2 and the 0.

The keyboard now finds itself in a cross-fire.

Knowing that your finger has the (human) kindness of continuing to press the key for more than a few milliseconds (even if you have very small fingers), we take advantage of this time to permute the initialization of the rows and columns by sending to the PCF 8574 (I²C write) a new value:

| LED | LINES | | | COLUMNS | | | |
|-----|-------|-----|-----|---------|-----|-----|-----|
| P7 | P6 | P5 | P4 | P3 | P2 | P1 | P0 |
| 1 | 1 | 1 | 1 | 1 | 0 | 0 | 0 |

and we run as fast as we can to reread the race results which are as follows (your finger being still on the eight, port P5 goes to 0):

| LED | LINES | | | COLUMNS | | | | |
|-----|-------|-----|-----|---------|-----|-----|-----|-----|
| P7 | P6 | P5 | P4 | P3 | P2 | P1 | P0 | |
| 1 | 1 | 1 | 1 | 1 | 0 | 0 | 0 | byte no. 2 |

which we gently bring back via the I²C (read) to the microcontroller.

Now we only have to process the two bytes no. 1 and no. 2 to prove that it was, in fact, the key '8' that was pressed.

This is accomplished by generating the logical OR (ORL) of bytes 1 and 2.

| LED | LINES | | | COLUMNS | | | | |
|-----|-----|-----|-----|-----|-----|-----|-----|------|
| P7 | P6 | P5 | P4 | P3 | P2 | P1 | P0 | |
| 1 | 0 | 0 | 0 | 0 | 1 | 0 | 1 | byte no. 1 |
| 1 | 1 | 0 | 1 | 1 | 0 | 0 | 0 | byte no. 2 |
| 1 | 1 | 0 | 1 | 1 | 1 | 0 | 1 | result of ORL |

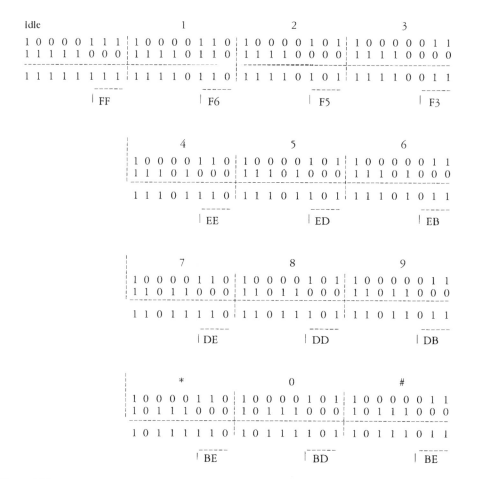

Figure 9.40

| Key | Results after 'logic OR' in hex values |
|-----|------------------------------------------|
| idle | FF |
| 1 | F6 |
| 2 | F5 |
| 3 | F3 |
| 4 | EE |
| 5 | ED |
| 6 | EB |
| 7 | DE |
| 8 | DD |
| 9 | DB |
| * | BE |
| 0 | BD |
| # | BB |

**Figure 9.41**

If you insist on understanding and if you have the fortitude, you can check all of the calculations of Figure 9.40. If not, you can go directly to Table 9.41 which gives the equivalence table of the codes obtained when each of the keys is depressed.

# The DISPLAY unit

## LTN 111 display unit

This module is made up of an LCD display and one or two integrated circuits.

One of them is a little microcontroller whose dedicated function is to manage the input data, put them in memory, associate them, via an integrated character generator (reconfigurable or not), with a precise meaning and to completely manage the display itself (multiplexing...).

The ways in which it can be used are too numerous to list here (if it is of any consolation, so are the numbers of errors that can be committed along the way), so we send you back to the manufacturer's documentation in case your applications are a bit special.

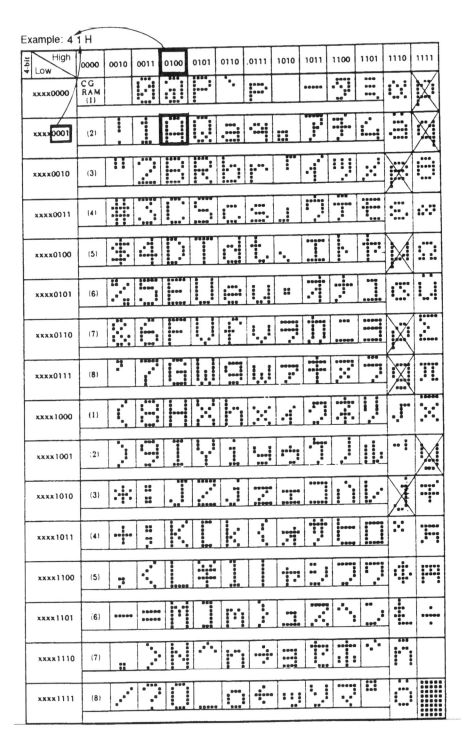

Figure 9.42

We have decided on moderation, at least to start with. We have chosen two circuits to control this module: one for the data, the other for control signals and the remote contrast control.

## Data

From D0 to D7 they correspond, more or less, to the ASCII codes! which substantially simplifies the software task and will certainly make your day. A good table is worth more than a long speech to describe all of the standard codes (Figure 9.42). Of course, the codes will have to be brought in. Never mind, they will take the bus like everyone else!

## Control signals

There is a timing consideration to be respected (Figure 9.43) and that's a bit more complicated. In addition, because of the possibility of a remote location of the keyboard/display unit, we have decided to let everything pass by the I²C bus. This leads us to use a trick, so that the signals will appear to vary correctly in time. Let's take a look at the way we have solved this frightful problem.

First, we send the data to the circuit IC₃ which latches them and keeps them on pins D0 to D7 of the display module. Obviously, nothing

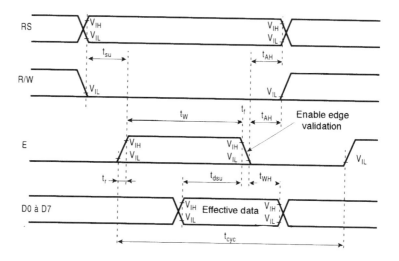

**Figure 9.43**

Initialisation of display module LTN 11R-100

Before initialisation first give values to PCF 8574:

| I²C slave addresses (PCF 8574) | |
| --- | --- |
| IC3 for data: 44h | IC4 for service signals: 42h |

Values of ports

| P7 | P6 | P5 | P4 | P3 | P2 | P1 | P0 | P7 | P6 | P5 | P4 | P3 | P2 | P1 | P0 |
| --- | --- | --- | --- | --- | --- | --- | --- | --- | --- | --- | --- | --- | --- | --- | --- |
| D7 | D6 | D5 | D4 | D3 | D2 | D1 | D0 | E | R/W | RS | - | -- | - | - |  |

STEP 1   Transmit via I2C bus

|  | | | | | | | | | value |  | | | | | | | | value |
| --- | --- | --- | --- | --- | --- | --- | --- | --- | --- | --- | --- | --- | --- | --- | --- | --- | --- |
| a) | 1 | 1 | 1 | 1 | 1 | 1 | 1 | 1 | FFH |  | | | | | | | |  |
| b) |  | | | | | | | |  | 1 | 1 | 1 | 1 | 1 | 1 | 1 | 1 | FFH |
| c) |  | | | | | | | |  | 1 | 0 | 0 | 0 | 0 | 0 | 0 | 0 | 80H |

STEP 2   Beginning of initialisation

| a) | 0 | 0 | 1 | 1 | 0 | 0 | 0 | 0 | 30H |  | | | | | | | |  |
| --- | --- | --- | --- | --- | --- | --- | --- | --- | --- | --- | --- | --- | --- | --- | --- | --- | --- |
| b) |  | | | | | | | |  | 1 | 0 | 0 | 0 | 0 | 0 | 0 | 0 | 80H |
| c) |  | | | | | | | |  | 0 | 0 | 0 | 0 | 0 | 0 | 0 | 0 | 00H |

STEPS 3 and 4   Repeat same sequence up to step 2

**Figure 9.44(a)**

happens on the display, which is politely waiting for its appropriate control signals. Secondly, given the special arrangement of pins that we have chosen for the IC₄ port of the PCF 8574 and those of the enable and read/write and the RS, we have cheated by sending a message, via the I²C bus, made up of two bytes whose only purpose is to make the display module think that, first, its enable is in the ONE state and, then, with the help of the second byte, slyly pass it to ZERO to obtain the fateful falling front which allows the validation of the orders necessary for triggering the display.

All of this is detailed in the table where we show how to display an 'A'. After that, you can have a lot of fun with the only remaining (big) detail. An initialization sequence is necessary for the display to operate.

STEP 5

| | | | | | | | | | | | | | | | | | | | |
|---|---|---|---|---|---|---|---|---|---|---|---|---|---|---|---|---|---|---|---|
| a) | 0 | 0 | 1 | 1 | 1 | 0 | 0 | 0 | 38H | | | | | | | | | | |
| b) | | | | | | | | | | 1 | 0 | 0 | 0 | 0 | 0 | 0 | 0 | 80H |
| c) | | | | | | | | | | 0 | 0 | 0 | 0 | 0 | 0 | 0 | 0 | 00H |
| d) | 0 | 0 | 0 | 0 | 1 | 0 | 0 | 0 | 08H | | | | | | | | | | |
| e) | | | | | | | | | | 1 | 0 | 0 | 0 | 0 | 0 | 0 | 0 | 80H |
| f) | | | | | | | | | | 0 | 0 | 0 | 0 | 0 | 0 | 0 | 0 | 00H |
| g) | 0 | 0 | 0 | 0 | 0 | 0 | 0 | 0 | 01H | | | | | | | | | | |
| h) | | | | | | | | | | 1 | 0 | 0 | 0 | 0 | 0 | 0 | 0 | 80H |
| i) | | | | | | | | | | 0 | 0 | 0 | 0 | 0 | 0 | 0 | 0 | 00H |
| j) | 0 | 0 | 0 | 0 | 0 | 1 | 1 | 0 | 06H | | | | | | | | | | |
| k) | | | | | | | | | | 1 | 0 | 0 | 0 | 0 | 0 | 0 | 0 | 80H |
| l) | | | | | | | | | | 0 | 0 | 0 | 0 | 0 | 0 | 0 | 0 | 00H |

display unit LTN 111

STEP 6  Beginning of display activities

| | | | | | | | | | | | | | | | | | | | |
|---|---|---|---|---|---|---|---|---|---|---|---|---|---|---|---|---|---|---|---|
| a) | 0 | 0 | 0 | 0 | 1 | 1 | 1 | 0 | 0EH | | | | | | | | | | |
| b) | | | | | | | | | | 1 | 0 | 0 | 0 | 0 | 0 | 0 | 0 | 80H |
| | | | | | | | | | | 0 | 0 | 0 | 0 | 0 | 0 | 0 | 0 | 00H |

and cursor appears

| | | | | | | | | | | | | | | | | | | | |
|---|---|---|---|---|---|---|---|---|---|---|---|---|---|---|---|---|---|---|---|
| c) | 0 | 1 | 0 | 0 | 0 | 0 | 0 | 1 | 41H | | | | | | | | | | |
| d) | | | | | | | | | then | 1 | 0 | 1 | 0 | 0 | 0 | 0 | 0 | A0H |
| e) | | | | | | | | | and | 0 | 0 | 1 | 0 | 0 | 0 | 0 | 0 | 20H |

and "A" appears

| | | | | | | | | | | | | | | | | | | | |
|---|---|---|---|---|---|---|---|---|---|---|---|---|---|---|---|---|---|---|---|
| f) | x | x | x | x | x | x | x | x | ??H | | | | | | | | | | |
| | | | | | | | | | | 1 | 0 | 1 | 0 | 0 | 0 | 0 | 0 | A0H |
| | | | | | | | | | | 0 | 0 | 1 | 0 | 0 | 0 | 0 | 0 | 20H |

and "A" appears

g)      etc     etc

**Figure 9.44(b)**

# INITIALISATION procedure

Two options are possible: a hardware option and a software option.

We find that the hardware method is a little dangerous (for your information, that is related to the power supply rise time constant!) and we prefer the sluggish but safe route. You will, therefore, suffer as much as we have by transmitting the initialization sequence proposed in Figure 9.44 before beginning the display of your charming messages.

# 10
# The Software

The purpose of this chapter is to describe examples of software which make it possible to operate the various microcontrollers whose hardware characteristics have been described in the preceding chapters. Before plunging into the twists and turns of this software, it is helpful first to examine the details of the organization of the internal memory spaces of these microcontrollers.

If you already happen to be familiar with the architecture of the 80 Cxxx family, we suggest that you go directly to the second part of this chapter, which deals with the software design of routines for the I²C bus.

# Part 1

# Architecture of the RAM and ROM Memories of the 8x Cxxx Family

We will now take a closer look at the architecture of the 80 C51 family, and show you how to organize your software in an orderly fashion. To accomplish this, we have to go back and spend some time on the internal memory structure of these microcontrollers, taking as an example the 80 C552. As we have already indicated, it is a direct offspring of the 80 C51 and, to be still more precise, of the 80 C52, due to the fact that its internal RAM has 256 bytes.

We will now ask you to look closely at Figures 10.1(a) and (b), which show the conventional configuration of the mapping architecture of the data memories (RAM), on the one hand, and the program memory ((EP)ROM), on the other hand.

We will try to guide you through this labyrinth of bytes, by beginning with the RAM, its makeup, its organization, its contents, its address modes and then, we will attack the (EP)ROM's before finally concluding with the overal memory space.

## The Memory Space of the 80 C51 Family

The addressing range of this CPU is, as a result of the architecture, two times 64 Kbytes, without using any special tricks. That means that this family is capable of addressing a total memory space of two times 64 Kbytes, which normally implies two address fields of 16 bits, plus 8 bits for data, or $(16 + 16 + 8) = 40$ output pins reserved for the 'address bus' and 'data bus'.

For very down to earth reasons, the number of pins has been intentionally reduced by declaring that the data memory space (RAM) and program memory space (ROM) share the same address pins. That is to say that, at the same digital value of the address, one can land either on RAM or ROM via a special addressing mode which we will discuss later (thanks to this principle, we have just saved sixteen pins!).

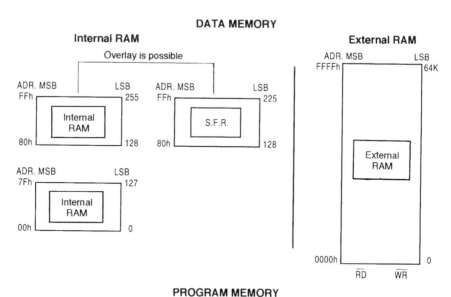

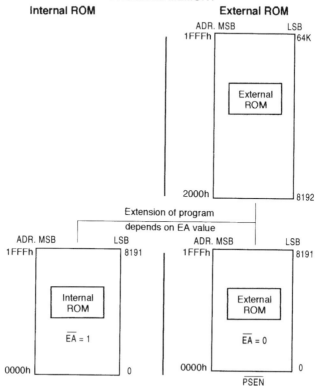

Figure 10.1

Another pin-saving method is the multiplexed Address and Data bus: we have split the 8 high order address bits (port 2) from the 8 low order bits (port 0), and time multiplexed the low order address bits and the data on the same pins in order to save an additional 8 pins (this is done because the cost of an external address latch circuit, necessary for demultiplexing; is cheaper when compared to the additional cost of the microcontroller pins). With that, we arrive at an overall total of $(8 + 8) = 16$ output pins for the total memory space (see Figure 10.2).

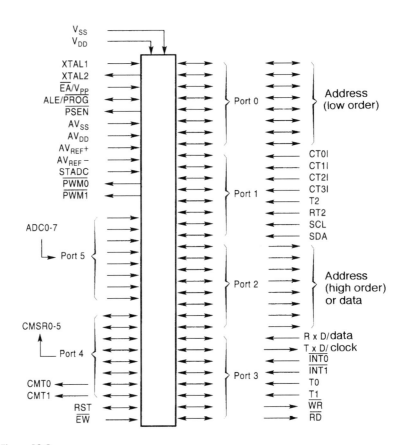

**Figure 10.2**

We have forgotten to mention two or three pins: it is necessary to indicate at what moment to latch the address register so as to be able to demultiplex the address/data bus. That is the function of the ALE signal (Address Latch Enable). The same is true for PSEN (Program Store Enable) and the R/W (Read/Write) which serve to indicate in which memory space one wants to work, and what type of operation we want to process.

All that makes, in fact, $(16 + 3) = 19$ pins instead of forty, but that's not so bad. Obviously, there will always be grouches who will tell you that they want a larger memory space and will talk about paging the memory field, deciding to use one or several bits of the input/ouput ports for this purpose. For the moment, we will not drag you to these distant lands from which certain interrupt vectors might have difficulty returning . . . . Everything in its time!

Finally, don't forget that it is possible, with the help of a wired AND, to superimpose the two RAM and ROM spaces in order to be able to execute code which would be intentionally located (temporarily or not) in the RAM (see Figure 10.3).

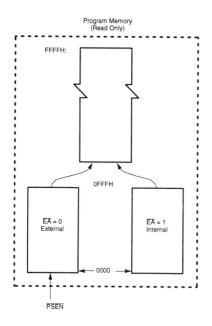 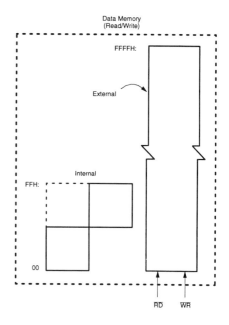

**Figure 10.3**

# The Data Memory

Its structure is very specific and deserves particular attention, especially in the case of the 8x C52 or its big brother 8x C552. This data memory is divided into several parts, each having a number of special features. Let us first start with a few comments on its general organization.

As shown in Figure 10.1, a part of this memory is internal and includes, in the case of the 80 C552 $(128 + 128 + 128)$ bytes; but, watch out, not all bytes are equal! In parallel with this internal memory, there can be, externally, a physical data memory space (RAM) of up to 64 Kbytes.

## The internal data memory

Let's take a magnifying glass and examine a little more closely how the inside of our microcontroller is organized. Figure 10.1 gives some complementary information. There, one sees two major blocks which have been designed for different purposes.

The very special right hand block, called 'Special Function Register', or SFR, which will be dealt with in great detail a bit further on, occupies addresses 128 to 255 (decimal) and the left hand block, which is the 'normal' part of the internal data memory. Let's look at this last part in more detail.

### The 'normal' part of the internal memory

Figure 10.4 shows its general skeleton, in which you can see various zones. How strange! At the expense of some momentary complications, it is intended to simplify your life, so have faith! You will notice that the lower part of this section of the data memory is cut into slices (four in number) called register banks (from 00 to 11, inclusive, in hexadecimal), each of which contains 8 bytes, rebaptized under the names R0 to R7. The hexadecimal addresses of each of them are indicated in the figure itself. One can read and write only in complete bytes in this part of the memory. These 4 banks of 8 registers take care of the first thirty-two bytes (from 00h to 1Fh).

Let's now look at their sixteen brothers located immediately above (from 20h to 2Fh). These bytes are absolutely unstructured and don't want to hear anything at all about bank organization! Moreover, each bit of each byte in this zone has can have an independent purpose with respect to its fellow bits in the byte where it lives. In more rigorous terms, it is a bit-by-bit addressable zone (as well as addressable in complete bytes). Up to you the maddening drudgery of setting, bit by bit, data bits that have nothing whatever to do with each other. Now let's come back to the other well behaved and disciplined bytes that live above (from 30h to FFh). Not the slightest idea of a data bank, not a single independent bit. Total monotony – simple routine!

You might think that everything is the same here – but it's not! As you will discover a little later, we have not yet told you everything. There

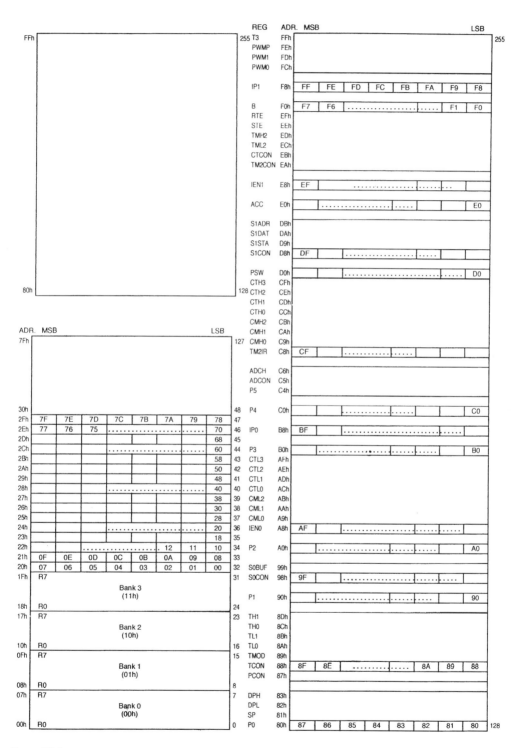

Figure 10.4

is a significance of the frontier at the level of 7Fh and 80h which we will explore later. After this outpouring of bytes, a little pause is in order to assure their effective digestion and then you will be up to tackling the SFR's!

## The SFR's of the internal data memory

Here is one of the crucial moments in the history of the 80 C51 and its derivatives. In the course of the design of the heart of the CPU, it was decided to set aside a particular memory space to allow the modification and/or upgrade of the operation of future versions of the microcontroller. A field of 128 bytes has been set aside for this purpose by the manufacturer. Figure 10.5 is a plan representation of this memory space.

| | | | | | | | | | |
|-----|------|------|-----|-----|-----|-----|---|------|-----|
| F8h | | | | | | | | | FFh |
| F0h | B | | | | | | | | F7h |
| E8h | | | | | | | | | EFh |
| E0h | ACC | | | | | | | | E7h |
| D8h | | | | | | | | | DFh |
| D0h | PSW | | | | | | | | D7h |
| C8h | | | | | | | | | CFh |
| C0h | | | | | | | | | C7h |
| B8h | IP | | | | | | | | BFh |
| B0h | P3 | | | | | | | | B7h |
| A8h | IE | | | | | | | | AFh |
| A0h | P2 | | | | | | | | A7h |
| 98h | SCON | SBUF | | | | | | | 9Fh |
| 90h | P1 | | | | | | | | 97h |
| 88h | TCON | TMOD | TL0 | TL1 | TH0 | TH1 | | | 8Fh |
| 80h | P0 | SP | DPL | DPH | | | | PCON | 87h |

Normal: bytes addressable registers
Bold: Bit addressable registers

**Figure 10.5**

As shown in the figure, the space is largely empty in the case of the basic microcontroller 80 C51 and leaves all kinds of openings for creating higher performance microcontrollers requiring supplementary controls.

It is notably the case of the 80 C552, where added A/D converters, PWM's, etc., necessitates complementary controls which flourish, like so many new SFR's, in the spaces available in this plan. This is shown in Figure 10.6. It should be noted that, as an aid to understanding, we have shown in bold the new registers belonging to the C552, and the corresponding registers of the basic microcontroller (80 C51). As you can see, this compatility allows the upward compatibility of software already developed on the basic microcontroller.

| ADR | | | | | | | | | ADR |
|-----|------|------|--------|-------|------|------|------|------|-----|
| F8h | IP1 | | | | PWM0 | PWM1 | PWMP | T3 | FFh |
| F0h | B | | | | | | | | F7h |
| E8h | IEN1 | | TM2CON | CTCON | TML2 | TMH2 | STE | RTE | EFh |
| E0h | ACC | | | | | | | | E7h |
| D8h | S1CON | SISTA | S1DAT | S1ADR | | | | | DFh |
| D0h | PSW | | | | | | | | D7h |
| C8h | TM2IR | CMH0 | CMH1 | CMH2 | CTH0 | CTH1 | CTH2 | CTH3 | CFh |
| C0h | P4 | | | | P5 | ADCON | ADCH | | C7h |
| B8h | IP0 | | | | | | | | BFh |
| B0h | P3 | | | | | | | | B7h |
| A8h | IE | CML0 | CML1 | CML2 | CTL0 | CTL1 | CTL2 | CTL3 | AFh |
| A0h | P2 | | | | | | | | A7h |
| 98h | SCON | SBUF | | | | | | | 9Fh |
| 90h | P1 | | | | | | | | 97h |
| 88h | TCON | TMOD | TL0 | TL1 | TH0 | TH1 | | | 8Fh |
| 80h | P0 | SP | DPL | DPH | | | | PCON | 87h |

Normal: Same registers as 80 C51
Italic: New or modified register
Bold: Addressable bits

**Figure 10.6**

Compatibility between family members is one of the key features of the 80 C51 family. This offers you the freedom of upgrading (or downgrading) to a derivative best adapted to your application, both in performance and in cost, and transport your existing software (even down to the junior versions of 80 C51 such as the 80 C751 and 80 C752). Contrary to what you might have thought, these SFR's are not all identically addressed. There are some which stand out above the others! In fact, all those in the extreme left hand column of Figure 10.6 (and, consequently, with addresses spaced in steps of eight and ending by 00 or 08h) are addressable bit by bit; all of their other comrades are only addressable by complete bytes.

Figure 10.7 summarizes the first 128 bytes of internal RAM spaces, bit addressable and byte addressable.

Before definitively leaving this section, let's give a tender thought to those poor component manufacturers who, in the course of the designing new 80 C51 derivatives, have to resolve very cruel dilemmas in deciding where to place new SFR's which will so simplify things to come! Whether they are bit by bit addressable or not, each SFR (byte) is composed of eight bits each having a precise meaning and action. Recognizing their individuality and in order to make them work well for us, to tame and train them, we have decided to identify each bit of each SFR with a specific name. As an example, take the SFR 'PSW' (Figure 10.8).

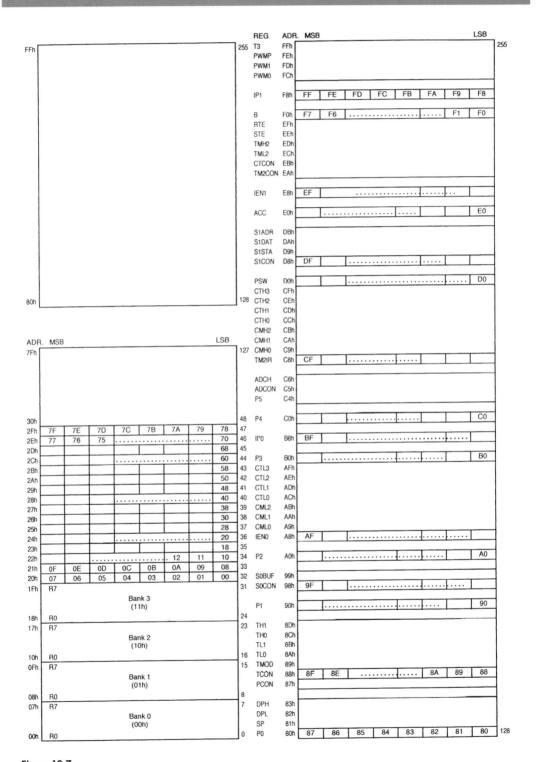

Figure 10.7

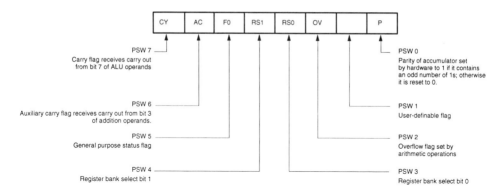

**Figure 10.8**

Figure 10.9 gives you the complete mapping (byte by byte, bit by bit) of all of the SFR's of the 80C552 that we can use in our applications (...to be learned by heart, of course!).

And now the scene is set. You know exactly who lives where in this memory space known as the internal data memory, but, unfortunately, lots of folks sometimes live at the same address! All we have to do to differentiate between the 'Normal' upper 128 bytes of memory and the SFRs is in the way we address them: indirect addressing always talks to the 'normal' memory, direct addressing always talks to the SFRs!

Figure 10.10 tells you much more than a long speech. You can see clearly that in the common address overlap space (from 80h to FFh), according to whether the addressing mode is direct or indirect, you can land either in the normal memory zone or in the SFR zone.

# The Program Memory

As previously indicated, the program memory can extend (normally) to 64 Kbytes, part of which can be resident in the microcontroller chip, in the case of applications where the program is written in internal ROM.

## Working hypothesis

With the object of simplifying our systems, we will tie the pin EA to +5 V and only operate in external program memory, that is, with the 80 Cxxx 'ROM-less' type and not the 83 Cxxx.

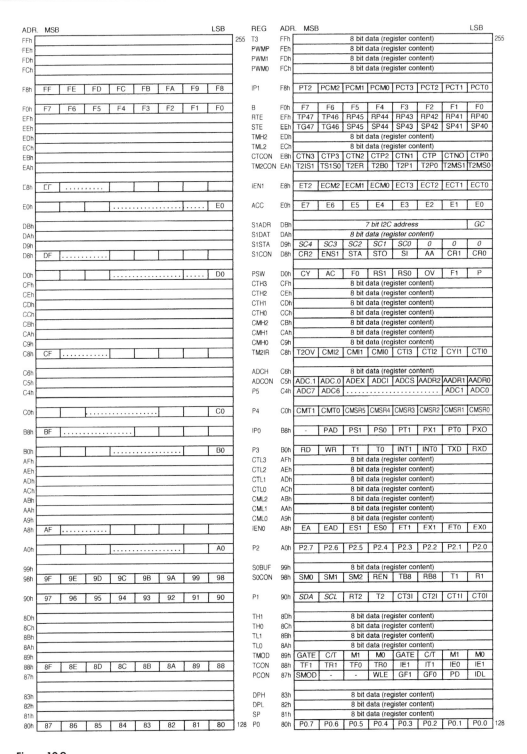

Figure 10.9

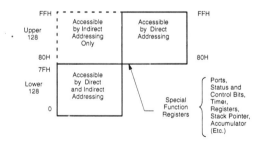

**Figure 10.10**

Note that, from now on, this will oblige us to use (or to condemn us to use) the port P0 for addressing and to go look for (by demultiplexing the A/D bus, using a 74LS573) the executable codes in the external program memory. (If, subsequently, you want to go to a more compact solution, and if you have supplementary ports available, nothing prevents you from using an 87 Cxxx OTP (or UV)).

# Organisation of the program memory

First of all, you can decide to do whatever your heart desires in that part of the memory. There are no formal rules concerning its use, but there are some things that work out better if you take a second look. By carefully reading the classical characteristics of the microcontrollers, you will discover other strange beings called interrupts and their associated vectors. Once you get over the shock, you very quickly realize that, despite the magnanimous statement that you can do whatever you like in the program memory space, you have been offered a slightly poisoned apple!

# Interrupts vectors

For an unlimited number of reasons, a microcontroller can and should be interrupted in the course of the execution of its main program (by the arrival of external event like termination of a count down, a button or key is pressed, or someone breaks into your living room etc.) by signals called interrupts. General alarm on board!

How do we process this interrupt and, above all where do we go to look for the beginning of the little piece of program which is allocated to this particular type of task, associated with this specific interrupt?

The microcontrollers have had a liberal education because, for quite some time, they know exactly that, for such and such a specific interrupt, it is necessary to go look in such and such a memory cell for the beginning of the corresponding software, or 'service routine'.

The address of this memory cell bears the lovely name of interrupt vector and, obviously, the greater the different interrupts available on the microcontroller, the greater the associated interrupt vectors.

## The interrupt vectors of the 80 C552

Figure 10.11 shows their functions, their names and the order in which they take the stage in the case of the 80 C552. Their numbers (let's be a little more serious: their addresses) are, nevertheless, special, if you look at them closely. They are all located at the beginning of the memory and they are spaced in groups of eight. All that did not happen by chance.

All of these interrupt vectors allow you to organize the overall management of your many and various interrupts. If your interrupt program is very short, it will fit easily in the seven bytes available to it. If not, with the help of the LCALL instruction, you can follow its progress anywhere that you see fit (or almost anywhere). You have probably noticed that we have also shown the vector 00h, which is the reset vector (hardware or semi-software) if you don't mind referring to this as a special interrupt. Now we have reserved the program memory from 00h to about 0FFh just for interrupt vectors and service routines!

The main program (yours, the one that you are going to develop) will certainly use sub-routines dedicated to repetitive tasks (for example: management of the I²C bus, A/D conversion, etc.). It is a good idea, as of now, to reserve space in the program memory in order to place these routines, at your leisure, in a well ordered fashion and, consequently, to start your main program somewhere in the higher levels of program memory (see the example in Figure 10.12) especially if you have room! After all of these generalities on the software organization, let's make a date for a discussion of the operation of the 'high performance' I²C bus hardware interface of the 80 C552, in the second part of this chapter, where we will help you to organize RAM and ROM so that the organization of your program memory will be clean and (even) readable by someone other than yourself!

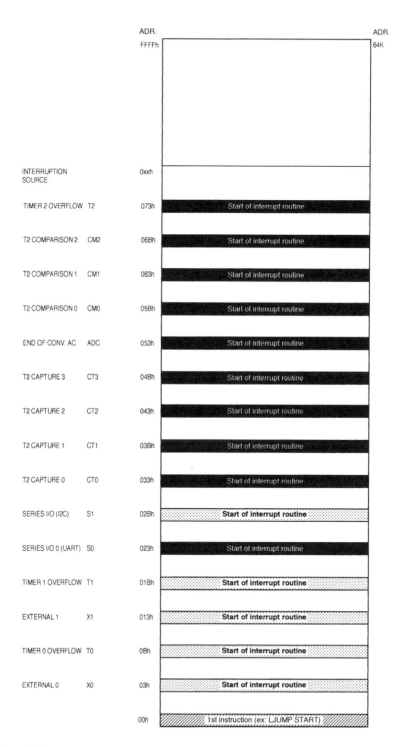

Figure 10.11

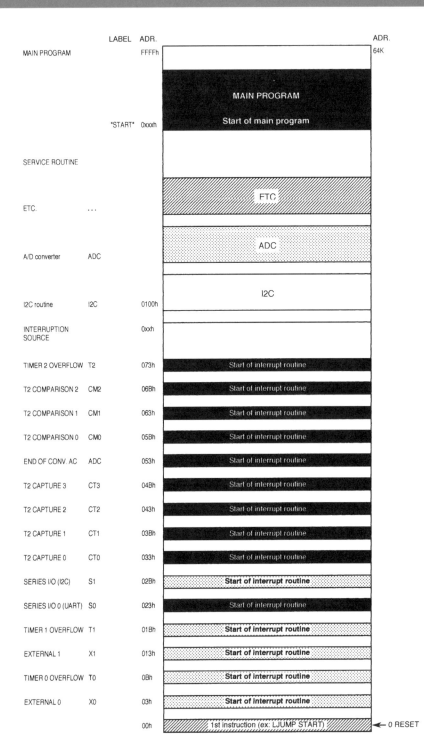

Figure 10.12

# Part 2

# Design of the I²C Routines

The purpose of this second part is to involve you in the conception of the control software for the I²C bus. First of all, we will tackle the problem using a microcontroller provided with integrated hardware I²C interface (example: 8x C652, C654, C552...). Then we will show you an example using an 80 C51 family microcontroller without I²C interface (using bit emulation of the I²C protocol, or 'bit-banging').

## Software for a Microcontroller with a Hardware I²C Interface

We are going to design I²C control software for microcontrollers containing a hardware I²C interface (example: the 8x C652, C654, C552). You will remember that we have already touched on this topic in the section concerning the register S1STAtus of the 8x C552. If not, you'd better review it.

### Organization of the software

We have included this software on the disk provided. The software is made up of five major blocks:

1. The declarations which only concern the assembler;

2. The initialization routines of this software module, which we can locate with its little initiating comrades wherever we see fit (in our case, to keep things clean, we have located them all on one page (page 2) of 256 bytes, especially created for that purpose, starting with ROM addresses 0200h);

3. A few lines of codes for communicating with the main program depending on the type of I²C peripheral we're talking to, number of bytes in the message, etc.;

4. The I²C interrupt management routine triggered by the SIO1 interrupt;

5. The steering routine for the twenty-six situations that may occur on the I²C bus, written on a bright and shiny new page: page 1 (100h and above, up to 1FFh). Figure 10.13 gives an overview of the locations of these routines.

Now let's look at their contents.

## The declarations

Declarations are very explicit and are carefully chosen to tell you the addresses of the Special Function Registers, SFR's. We have also defined the values with extended names, the page where we are going to work (which you can of course change if by chance it is already occupied) and, finally, the names and addresses of the RAM registers which we have requisitioned to make the routine run.

## Initialization routine

We have decided that, after the general reset of the microcontroller (which resets the program counter to address zero), we will begin by sending the software out to make the rounds of the initialization routines located at 0200h. This routine first initializes the SIO1 interface to be either a slave-transmitter or a slave-receiver, designed so as to be always ready to listen to the I²C bus.

You might tell us that there is only one microcontroller on your card and that all this seems wasteful; to which we would reply that often a microcontroller all alone gets bored and, sooner or later, feels the need to receive messages from one of its buddies and that to save three bytes is just not worth it if it deprives you of that possibility. And, that's the way it is. But don't worry, our micro will very soon become a master-transmitter or master-receiver.

## Starting the exchange

We have decided as an example to transmit initially or receive 4 bytes, as master of the exchange. It is first necessary to write something in the register NUMBYTMST (number of the bytes to be sent or received).

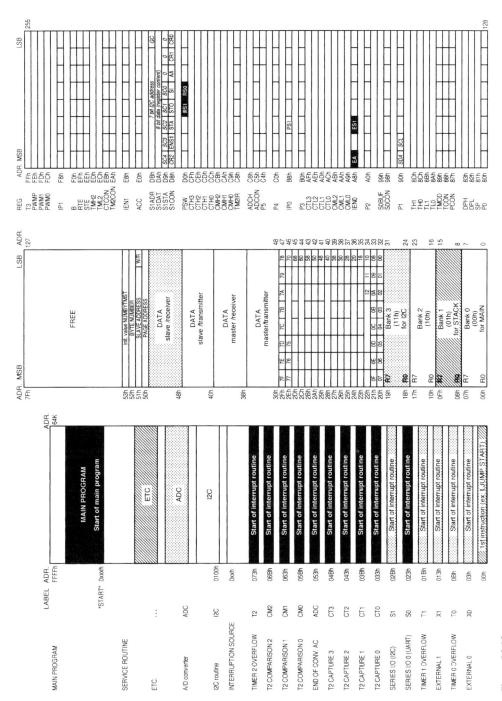

**Figure 10.13**

Obviously, if you want to load an 8-bit peripheral port such as the PCF8574, one data byte is enough. It is up to you to write these charming little lines (three, in all – big deal!) as a function of the type of circuit that you want to address. The last line of this mini-program sets the bit STA(rt) in the S1CON register and, all aboard, we're ready to go: the START condition is generated on the bus.

Let's go now to the main course: the interrupt routine of the SIO1 interface. What a tasty morsel!

## The interrrrupt routine

This routine is composed of 7 bytes and is accomplished in 8 machine cycles (about 7 or 8 $\mu$s). Let's look in detail at how it has been designed.

The START condition has just taken place. The SIO1 interface, which systematically rereads the bus level to check what's going on, registers that the START condition has actually occured on the bus. The interface now loads the value of 08h in the S1STA register, which leads to the next series of events...

The microcontroller goes into an interrupt state! Don't panic! Keep in mind these three points:

1. The SIO1 interrupt behaves like a hardware-generated LCALL instruction whose purpose is to call a sub-routine from wherever you are in the program. It will then call the I²C interrupt vector called S1, one of those that we presented in detail previously. This vector normally lives at the address 02Bh of the program memory.

2. The microcontroller, not knowing much about what we are going to do, saves the value of the program counter (two bytes) by PUSH(ing) it onto the stack (internal RAM space whose initial location you will have previously defined by setting the stack pointer ), saying to itself that after the interrupt, it can at least go back to whatever it was doing,and know where it had been. That is, it will POP (or else STACK) the PUSHes (see Figures 10.14 and 10.15).

3. Now the micro waits for your carefully written instructions. The ball is in your court. Standard good manners require you to save the program status word PSW which tells the microcontroller in which register bank it was working. Returning from the interrupt, the microcontroller can POP the PSW to start out again on the right foot and avoid mixing apples with pears in your magnificent program.

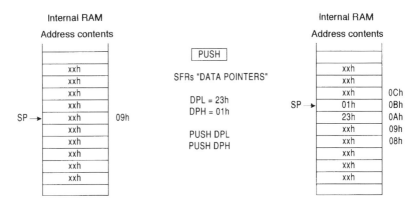

**Figure 10.14**

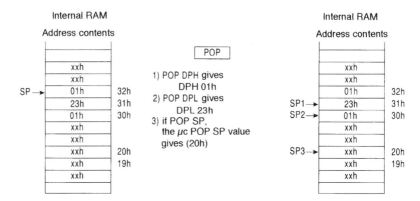

**Figure 10.15**

We have just finished with the first line of the interrupt routine! Fortunately, there are only four of them!

Now for a little trick.... For no extra charge, let's PUSH, in order (that's very important), S1STA (which holds the I²C status information) and then HADD (which holds the value of the page where the interrupt vectors are located). Don't try to understand...yet!

And now, a little touch of assembly instruction RET. It is quite simple, but you can't imagine what it is going to lead to. The purpose of the RET(urn) instruction is to come back from the 'false' LCALL and, therefore, to POP the number of bytes that had been PUSHed, that is to say, two. If, in the mean time, nobody has played with the PUSHes and the POPs, it will come back up with those bytes that it had put there and will simply attribute them to the high and low bytes of the PC (program counter), so that we can go on from there. However, in PUSHing S1STA, then HADD without intentionally unPOPping them, we have sown the seeds of discord in this wonderful machine, all the better to confuse it!

The RET instruction will, therefore, POP the two last values entered and attribute them to the Program Counter and the result will be:

|  | PC high = HADD which, here, is | 01h |  |
|---|---|---|---|
| and | PC low = S1STA which is | | 08h |
| so that | PC = HADD, S1STA = | 01 | 08h |

and the program will 'stupidly' continue from the program memory address 0108h. So what do we have here now? The routine which corresponds to the value contained in S1STA, whatever it is!

By solving this problem, we have just (re-)invented auto-vectoring of the interrupt routine. This will allow us to properly manage the myriad of conditions that could occur on the I²C bus. Clever, isn't it?

## The steering routines

Now come running the various cases that can come up in the course of the exchange, under different master/slave or transmitter/receiver conditions. There is nothing special to point out in their regard because everything usually fits in the 8 fateful bytes of status register spacing. This allows a clear layout of the software, thanks to which you will have a solid and reliable connection via this I²C hardware and software interface.

You can find the programs, both in assembly and C language, on floppy disk.

# Software for a Microcontroller Without a Hardware I²C Interface

The purpose of this section is to describe how to create simple software which will allow communication according to the I²C protocol for an 80 C51 family microcontroller without integrated I²C interface.

The example described will make it possible to communicate with a module equipped with the Philips SAA1064 LED driver described earlier. Those who would like to use this type of bus on other types of microcontrollers can use the flowchart of this program. It is of course important to adapt the timing relationships of the program execution speed to the timings required by the I²C protocol.

Before really beginning to examine this software example, let us take a look at the type of signals that we would like to obtain.

# Timing considerations for the I²C bus

The I²C bus protocol, in its basic structure, has been thoroughly described in the preceding chapters so we are not going to go back to it again. Without further ado, we will create software to manage it. In addition to the relations and conventions of the protocol, it is necessary to keep in step with the I²C's rather special timing specifications (we will work here with 100 kbit/sec I²C). Figure 10.16 shows the important I²C timing specifications which must be obeyed.

Figure 10.16 shows principal I²C timing values to keep in mind. Three of them need to be examined with very special attention:

| Parameter | Symbol | Standard-mode I²C-bus | | Fast-mode I²C-bus | | Unit |
|---|---|---|---|---|---|---|
| | | Min. | Max. | Min. | Max. | |
| SCL clock frequency | $f_{SCL}$ | 0 | 100 | 0 | 400 | kHz |
| Bus free time between a STOP and START condition | $t_{BUF}$ | 4.7 | - | 1.3 | - | µs |
| Hold time (repeated) START condition. After this period, the first clock pulse is generated | $t_{HD;STA}$ | 4.0 | - | 0.6 | - | µs |
| LOW period of the SCL clock | $t_{LOW}$ | 4.7 | - | 1.3 | - | µs |
| HIGH period of the SCL clock | $t_{HIGH}$ | 4.0 | - | 0.6 | - | µs |
| Set-up time for a repeated START condition | $t_{SU;STA}$ | 4.7 | - | 0.6 | - | µs |
| Data hold time: for CBUS compatible masters (see NOTE, Section 9.1.3) for I²C-bus devices | $t_{HD;DAT}$ | 5.0 $0^{1)}$ | - - | - $0^{1)}$ | - $0.9^{2)}$ | µs µs |
| Data set-up time | $t_{SU;DAT}$ | 250 | - | $100^{3)}$ | - | ns |
| Rise time of both SDA and SCL signals | $t_R$ | - | 1000 | $20 + 0.1C_b^{4)}$ | 300 | ns |
| Fall time of both SDA and SCL signals | $t_F$ | - | 300 | $20 + 0.1C_b^{4)}$ | 300 | ns |
| Set-up time for STOP condition | $t_{SU;STO}$ | 4.0 | - | 0.6 | - | µs |
| Capacitive load for each bus line | $C_b$ | - | 400 | - | 400 | pF |

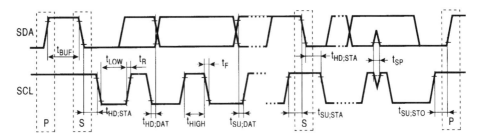

Figure 10.16

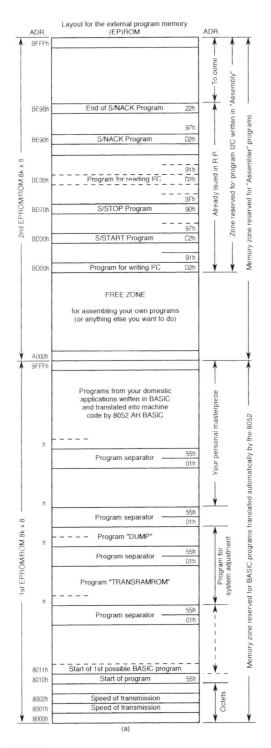

Figure 10.17 (a)

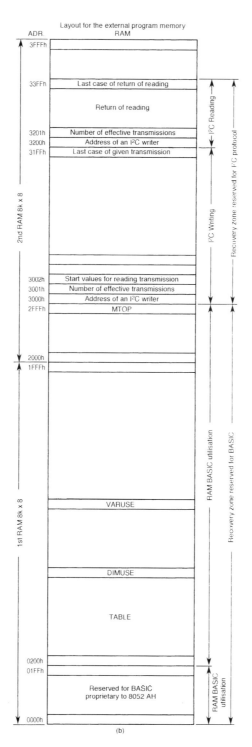

**Figure 10.17 (b)**

1. the start and stop conditions

2. the timing relationships between SDA and SCL for valid data transmission

3. the maximum operating frequency in standard mode (100 Kbits/s), if the bus is to function correctly

For our example that we have chosen the SAA1064 LED driver and the following messsage:

76h for the component address

00h for the instruction byte

77h for the control byte

For more detailied technical information on the SAA1064, we refer you to the preceding chapters where this component has already been dissected.

# Write routine for an $I^2C$ component

This program is designed to be nothing but a sub-routine which can be called as often as necessary by the main program whenever the LED needs to be updated. The parameters of this $I^2C$ write routine are organized as shown in Figure 10.17 as:

the address of the $I^2C$ component contained in                          3000h

the number of bytes to be transmitted contained in                       3001h

their respective values of the data to be transmitted beginning at 3002h

The only output parameter that needs to be sent by this assembly language program back to the main program is the value of the last $I^2C$ slave acknowledgement (at address 3FFEh).

The addresses just mentioned are, of course, addresses of the external RAM that we have decided to reserve for this particular purpose. The main program is, in itself, quite general:

♦ it initializes the various $I^2C$-related registers,

♦ it loads the register DPTR with the addresses containing the bytes to be transmitted,

♦ decrements the contents of the DPTR, so long as there is something to transmit or the receiving component does not interrupt the transmission with a non-acknowledgement.

Let's look at this program in a little more detail (Figure 10.18).

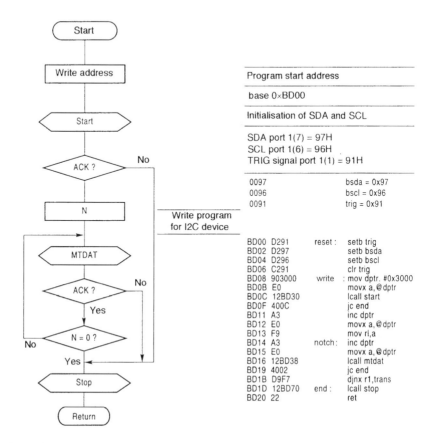

**Figure 10.18**

Lines 1 to 4 serve for initialization, after which the value 3000h is loaded into the DPTR and its contents is loaded into A (starting from the label WRITE), before calling the sub-routine START. The START routine's purpose is to to carry out the start procedure and to wake up the receiving component.

Line 8 allows testing to see if there is an acknowledgement (carry set to zero). If not, the receiver did not recognize the address and the program stops. This would happen, for instance, if the SAA1064 were not-connected, or non-functioning.

As we have seen, the DPTR is then loaded with the address 3001h, containing the number of bytes to be sent. This value will be stored in the register R1 which will be used as a down counter. As long as R1 is different from zero, the program will loop on the label TRANS, thus allowing the transfer of all of the bytes via the call MTDAT.

When it has finished, a STOP is generated and control is passed back to the main program.

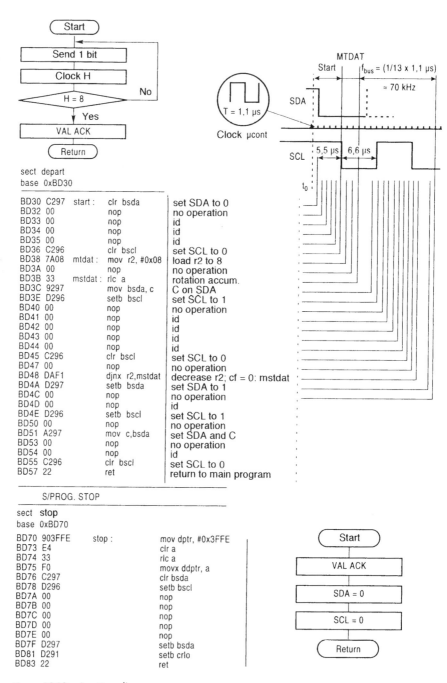

```
sect  depart
base  0xBD30

BD30  C297   start :    clr bsda          set SDA to 0
BD32  00                nop               no operation
BD33  00                nop               id
BD34  00                nop               id
BD35  00                nop               id
BD36  C296              clr bscl          set SCL to 0
BD38  7A08   mtdat :    mov r2, #0x08     load r2 to 8
BD3A  00                nop               no operation
BD3B  33     mstdat :   rlc a             rotation accum.
BD3C  9297              mov bsda, c       C on SDA
BD3E  D296              setb bscl         set SCL to 1
BD40  00                nop               no operation
BD41  00                nop               id
BD42  00                nop               id
BD43  00                nop               id
BD44  00                nop               id
BD45  C296              clr bscl          set SCL to 0
BD47  00                nop               no operation
BD48  DAF1              djnx r2,mstdat    decrease r2; cf = 0: mstdat
BD4A  D297              setb bsda         set SDA to 1
BD4C  00                nop               no operation
BD4D  00                nop               id
BD4E  D296              setb bscl         set SCL to 1
BD50  00                nop               no operation
BD51  A297              mov c,bsda        set SDA and C
BD53  00                nop               no operation
BD54  00                nop               id
BD55  C296              clr bscl          set SCL to 0
BD57  22                ret               return to main program
```

S/PROG. STOP

```
sect  stop
base  0xBD70

BD70  903FFE     stop :     mov dptr, #0x3FFE
BD73  E4                    clr a
BD74  33                    rlc a
BD75  F0                    movx ddptr, a
BD76  C297                  clr bsda
BD78  D296                  setb bscl
BD7A  00                    nop
BD7B  00                    nop
BD7C  00                    nop
BD7D  00                    nop
BD7E  00                    nop
BD7F  D297                  setb bsda
BD81  D291                  setb crlo
BD83  22                    ret
```

Figure 10.18    (continued)

**(a)**

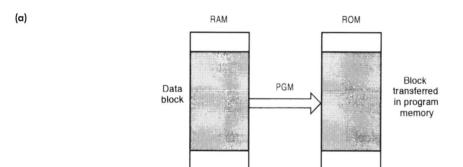

**(b)**

```
5       MTOP=019FFH
10      REM PROG. TRANSRAMROM
15      ADR=0 : COD=0
20      PRINT  :  INPUT "ADRESSE DE TRAVAIL ?",ADR
25      S=ADR :  IF S<=MTOP.OR.S>3FFFH THEN 15
30      DO
35      PRINT  :  INPUT "VALEUR DU CODE EN HEXA ?",CODE
40      XBY(ADR)=CODE
60      ADR=ADR+1
65      WHILE CODE<255
70      E=ADR-2 :  IF E<S.OR.E>3FFFH THEN 15
75      PRINT "TYPE DE MEMOIRE:"
80      PRINT "  1=EEPROM"
85      PRINT " 2=INT EPROM"
90      PRINT "3=NORMAL EPROM"
95      PRINT  :  INPUT "TYPE (1,2,3) ?",T
100     ON (T-1) GOSUB 310,315,320
105     IF W=.001 THEN DBY(38)=DBY(38).OR.8 ELSE DBY(38)=DBY(38).AND.0F7H
110     PUSH (65536-(W*XTAL/12))
115     GOSUB 330
120     POP G1
125     DBY(40H)=G1
130     POP G1
135     DBY(41H)=G1
140     PRINT  :  INPUT "ADRESSE DE DEBUT DE STOCKAGE EN ROM ?-",P
145     IF P<8000H.OR.P>0BFFFH THEN 140
150     PUSH (E-S)+1
155     GOSUB 330
160     POP G1
165     DBY(31)=G1
170     POP G1
175     DBY(30)=G1
180     PUSH (P-1)
185     GOSUB 330
190     POP G1
195     DBY(26)=G1
200     POP G1
205     DBY(24)=G1
210     PUSH S
215     GOSUB 330
220     POP G1
225     DBY(27)=G1
230     POP G1
235     DBY(25)=G1
240     PRINT  :  PRINT "TAPER 'CR' POUR CONTINUER"
245     X=GET :  IF X<>0DH THEN 245
250     PGM
255     REM CONTROLE DU BON TRANSFERT
260     IF (DBY(30).OR.DBY(31))=0 THEN  PRINT "FINDETRANSFERT" : END
265     PRINT "ERREUR" :  PRINT
270     REM SS PRG DE CALCUL DES ADRESSES INCORRECTES
275     S1=DBY(25)+256*DBY(27)
280     S1=S1-1
285     D1=DBY(24)+256*DBY(26)
290     print "LA VALEUR ",XBY(S1), : print " A ETE LUE A L'ADRESSE ",S1
295     PRINT
300     print "LA LECTURE DE L'EPROM DONNE ",XBY(D1), : print" A L'ADRESSE ",D1
301 end
305     REM SS-PRG DONNANT LA VALEUR DE W
310     W=.0005 :  RETURN
315     W=.001 :  RETURN
320     W=.05 :  RETURN
325     REM SS PRG STOCKANT LE BIT DE POIDS FORT ET FAIBLE DE G1 EN PILE
330     POP G1 :  PUSH (G1.AND.0FFH)
335     PUSH (INT(G1/256))
340     RETURN
```

Figure 10.19   (English version available on program disk)

## Test program for the SAA1064

Now that we have our basic program flow, it is simply a matter of loading the correct values into our RAM before calling up routine. Suppose we want to see the numbers '1', '2', '3', and '4' on the LEDs. For our own convenience, we could assign names to each location based on it's function, for example:

ADRCOMP (=3000h)   the address of the I²C component we want to write to

ADRNBR (=3001h)   the number of bytes to be transmitted on the bus

ADRDEB (=3002h)   starting address of the bytes to be transmitted

In the case of the SAA1064, this gives the following:

(ADRCOMP)   = 76h (slave address)

(ADRNBR)   = 06h (number of bytes to be sent)

(ADRDEB)   = 00h (instruction byte)

(ADRDEB +1) = 77h (control byte)

(ADRDEB +2) = 48h   '1'

(ADRDEB +3) = 3Eh   '2'

(ADRDEB +4) = 6Eh   '3'

(ADRDEB +5) = 4Bh   '4'

You may recognize that the display of the four first digits 1, 2, 3 and 4 are values we have borrowed from the preceding chapters.

**SECTION 4**

# EXTENSIONS, BRIDGES AND TOOLS

# 11

# Extensions of the I$^2$C Bus

We are now going to show you how to escape from the mother board's apron strings! Some of the electronic escape modes available to you are:

+ using cable extensions

+ using radio frequencies

+ using infrared (IR) links.

## Extensions

Extension in other systems doesn't always mean flexibility. With the I$^2$C bus, however, extension certainly does offer you flexibility.

One of the I$^2$C bus great strengths is that each circuit has a specific address that can answer to its own name via an acknowledgment. This allows the user to design software that can handle many peripherals in many different configurations, and simplifies the troubleshooting procedure for each IC. This architecture also makes it easy to detect malfunctions and even to detect which peripherals are present on the bus (and which are not). This is particularly useful when we want to plug in an I$^2$C 'extension'. Provisions for these add-on 'accessories' can easily be accommodated in the software.

For example in a previous chapter we demonstrated a keyboard/display system. This is an example of an I$^2$C bus extension which may not be permanantly connected.

In fact, the I$^2$C bus extensions can be connected whenever it strikes your fancy, using a simple I$^2$C connector. If you have designed your software to include a periodic keyboard/display call, the system can recognize automatically whether the keyboard is physically connected, and vector to the keyboard monitoring software if it is. As soon as you disconnect the keyboard/display module, the system skips all of these routines. All through the 'magic' of the I²C acknowledge!

# A Thousand Ways to Extend the I²C Bus

Bus extension, in terms of larger distances, is always a controversial subject because everyone sees the possibility of performance degradation. So, how can we overcome degradation of I²C bus performance when we try to extend it?

Let's start at the beginning. This bus is asymmetrical with respect to ground. In addition, its specifications (*see the chapter on protocol*) tell us that two hardware parameters govern extension performance:

♦ the maximum current load capacity of the output stage

♦ the bus signals' rise time (the usual microsecond in standard mode).

In addition, a software parameter requires I²C to be bi-directional, which implies that the receiver must send an acknowledgment signal to the transmitter.

Now that the scene is set, we can now try to limit the possibilities.

# Wired extension of the I²C bus (first option)

### As is

Seemingly trivial problems can cause serious trouble! We already mentioned that the I²C bus manages quite well (the capacitance of wires and connected circuits included) with its maximum capacitive load of 400 pF. This usually translates to several metres of bus length (more than enough for most applications).

If we want to go even longer (translating to too high a bus capacitance...), it is necessary to reduce the source impedance that limits the I²C line to satisfy the rise time criterion.

### Buffering

The solution is simple: buffering the I²C bus bi-directionally! You can look at this process in two entirely different ways: as logic or as analogue.

## The logic solution

Figure 11.1 illustrates a circuit that works better in theory than in practice. Let's look at the circuit diagram. Theory shows that the function to be accomplished is satisfied. When you try to put it into operation, things take a turn for the worse.

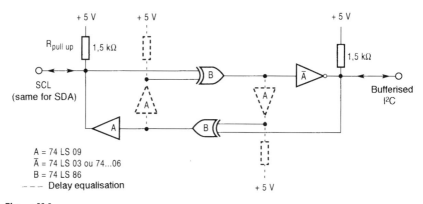

**Figure 11.1**

You already know that there is always a delay between a logic gate's output signal and its input, due to propogation delay. Because of this delay, the output signal operates as an input signal on the system output. By trying to assure bi-directionality, we have invented an oscillator. (If you don't believe it, give it a try!)

You must balance (or compensate for) all of the delays present in the loop by playing with each gates' nominal delay. Of course, each gate may represent different technologies and tolerances. Figure 11.2 illustrates some successful solutions. The main constraint is that these solutions lock you into using the same types of components in the same places.

It's the eternal engineering problem of a beautiful theory that doesn't work well in reality.

## The analogue solution

The circuit illustrated in Figure 11.3 may be less sophisticated, but it is quite effective.

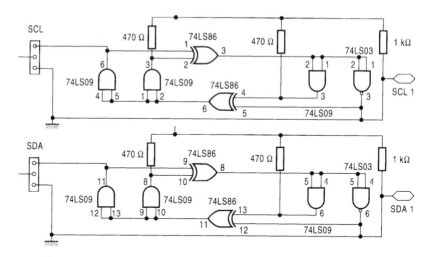

**Figure 11.2**

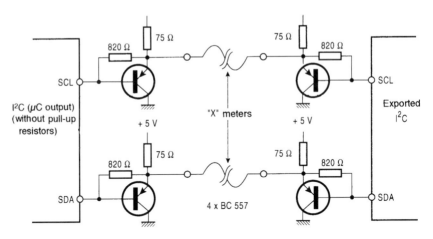

**Figure 11.3**

## How does it work?

We replaced the pull-up resistances to +5 V by PNP transistors in the emitter follower configuration, loaded by 75 Ω. When the SDA and SCL pins are at rest, T1 and T2 do not conduct and everybody is at +5 V.

Suppose that IC1 is master at the moment. T1 its emitter varies with the rhythm of the information. T2's emitter also varies. To prevent its base from doing whatever it wants, we took the precaution of adding an 820 Ω resistance. This forces it to follow the emitter's movement , controlling the input of the integrated circuit in question. Since the system is entirely symmetrical, similar reasoning applies in the other direction.

Are the bus electrical levels degraded or not? For a +5 V power supply, the bus specifications are:

in **input mode**:

Vil max = 1.5 V (maximum value of the input voltage in the **low** state);

Vih max = 3.0 V (minimum value of the input voltage in the **high** state).

in **output mode**:

Vol max = 0.4 V (maximum value of the output voltage).

There are no electrical problems because, in worst case we have:

Vol max + Vbe max = 0.4 + 0.8 = 1.2 V.

There is no problem at the high level due to the electrical configuration; the only degradation that can be criticized is a decrease in noise immunity that is largely compensated by the decrease in the impedance at which the signal is conveyed.


## Maximum distance

For long-distance transmission where the propagation time is the same or greater than the signal rise time, it is no longer appropriate to think of the line as a lumped capacitance. It is a true transmission line whose governing property is its characteristic impedance. Typical values of characteristic impedance normally fall between 75 and 100 Ω. The emitter follower drivers in Figure 11.3 can easily deliver the 100 mA or so required to drive a 75 Ω line, terminated at both ends, without requiring excessive base current from a standard microcontroller's I²C drivers. In the absence of excessive attenuation, the maximum transmission distance is limited only by the allowable propagation delay time, which is itself determined by the time that the transmitter can wait for the return of an acknowledgment.

The protocol stipulates that SCL clock pulses should remain in the high state for a minimum of 4.7 microseconds. The down and back propagation time of the line must be less than this value. Most transmission lines have propagation velocities of about 200 metres per microsecond so that, at least theoretically, a 400 metre line with a down and back propagation time of 4 microseconds would be possible with thick wires to limit the attenuation. Transmission distances of 100 metres and more remain well within the realm of possibility; this should be quite sufficient for most of your applications.

Although data is not rigorously protected in the encoding (parity, CRC...) the signals, though asymmetrical, are at quite low impedance and are relatively immune to noise.

### Realizing an integrated circuit

We have presented **discrete component** solutions for I²C buffering. Why isn't there an integrated circuit to do all that? There is! It's called the Philips' P82B715 bus buffer IC.

There are buffering solutions. Figure 11.4 shows the structure of 1/2 the P82B715 integrated circuit.

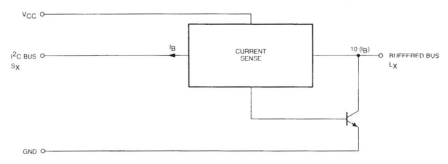

**Figure 11.4**

If the **buffer** circuit is permanently connected to an existing I²C circuit, the pull-up resistances present on the standard outputs should be removed and mounted directly on the buffer output (see Figure 11.5. For an application example, see Figure 11.6).

## From asymmetric to symmetric (second option)

We have lengthened the distance over which the bus runs, but we have not improved the integrity of the data conveyed or decreased degradation by external *parasitics*. We have substantially decreased line impedance, rendering it more immune to perturbing signals.

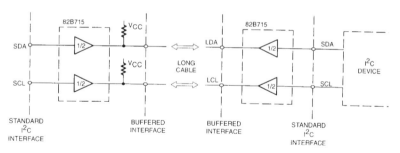

**Figure 11.5**

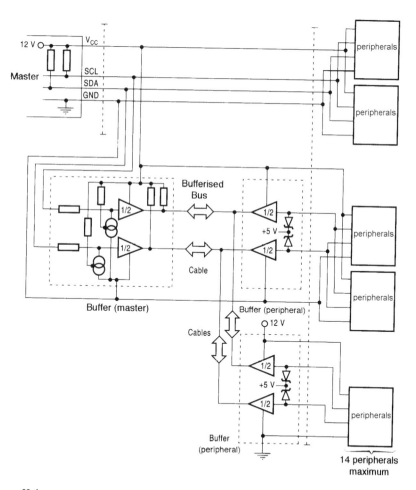

**Figure 11.6**

To combat *parasitics*, consider lines that are symmetrical, differential (and, of course, bi-directional). Figure 11.7 shows how you can achieve this. Watch out, because looping two operational amplifiers on themselves is a form of electronic suicide!

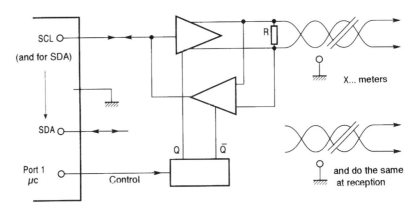

**Figure 11.7**

The remedy consists of controlling the coming and going of information at the right time by disconnecting the unused route with the help of an additional control signal. This is simple when the bus is software emulated; you can always output information on an additional (port) pin to control the direction of the exchange. It is more complicated to manage a hardware I²C bus interface because of the absence of a dedicated data direction indicator.

## Physical separation

We come to a turning point in the I²C story. This solution is our first effort to electrically separate the transmitter(s) from the receiver(s).

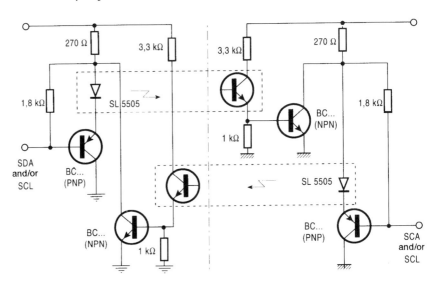

**Figure 11.8 (a)**

We are going to electrically separate the transmitter(s) from the receiver(s) while maintaining a tranmission link via infrared. This may be required by regulations or safety precautions when connecting pieces of equipment that do not physically use the same power supply. This is also used to connect systems working on 5 V and 3 V logic levels.

The setups in Figure 11.8(a) and (b) show the circuit diagrams we chose for IR mono-master and multi-master solutions. (They are only IR isolated adaptations of the preceding figure).

This setup is simple except for the choice of optocoupler, because of the one microsecond rise time that has to be respected independent of the binary data rate. You must use a high-speed optocoupler (such as the 6N139) to match the bus timing specifications.

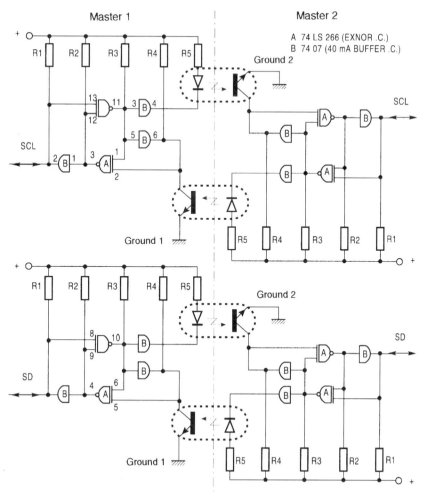

**Figure 11.8 (b)**

When you lay out your printed circuits for this solution, keep your choice of optocoupler in mind, as well as the distance that must be maintained between the primary and secondary pins (4 or 8 mm) to conform to class 1 or 2 isolation standards.

The block diagram in Figure 11.9 can provide a solution for multiple separations. In this case the transmitting diodes must be connected in series.

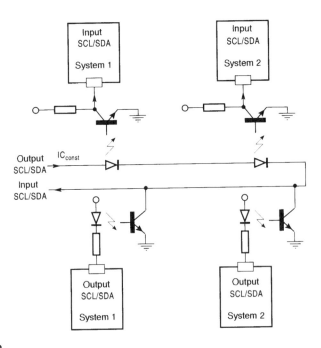

**Figure 11.9**

## Cutting the hard-wired connections (the third option)

Let's take an objective look at the realities in leaving wires behind. Among the means of transmission media available to us are radio frequencies (RF) and infra-red (IR).

IR transmissions practically imply a relatively short transmission distance because the transmitter and receiver must be in direct view of each other (in the same room). IR is mostly used in human-controlled functions which give immediate visual or audible feedback. In this case I²C can be easily implemented unidirectionally (acknowledgement can be ignored), and in master transmitter (write-only) mode.

For greater distances, RF can be used to send remote control messages thereby eliminating the direct visual contact requirement between the transmitter and the receiver. However, restrictions still apply.

Whether in IR or RF, we would have to transmit the two SDA and SCL signals simultaneously on two different channels (frequencies), and if we want to fully implement bidirectional communication and all the features of I²C, each channel would have to be split again into two channels each ... not an attractive solution, nor an easy one to design (although it has been done).

## I²C infra-red link

Instead of trying to fit a round peg in a square hole, it's often better to use a solution already optimized to solve the problem. Using commercially established IR remote transceivers with I²C allows many simplifications because off-the-shelf integrated circuits which satisfy most of our requirements exist already.

Our recommendation is an RC5 remote control transcoding that is extensively used for television and video tape recorders. It is a fifth-generation remote control that allows you to address 2048 previously defined or reconfigurable functions (see the chapter on bridges).

We propose a set-up that transcodes the RC5 code transmitted by the IR into I²C without acknowledging the transmission.

This hardware solution avoids complicated multi-channel encoders and decoders, and high-level software gymnastics that would be required to do straight I²C over-the-air. All you need to do is to choose the RC5 codes that you want to use in your own application.

A last remark: there is no direct relation between the flow rate of the RC5 code and the I²C bus, which continues to run over the range of zero to 100 kHz.

# 12 Bridges

## Part 1

## Bridges Towards Other Serial Buses

Now we come to a new stage in the social life of the I²C. So far we have talked about how the baby lived comfortably in the family home, then, as an adolescent, was allowed a few escapades outside and now, we arrive at a period of maturity where it tries to communicate with other protocols of the same species. Obviously, communication is not always easy when everyone has his own specific language. To get around this problem, it is often necessary to have recourse to coders/decoders

This leads us to bring up the question of bridges, applicable both to software and to hardware. To avoid having to write an encyclopedic volume describing the almost unlimited number of existing software bridges (often using a great deal of CPU time), we have chosen to indicate only the few hardware bridges most frequently used in association with the I²C bus:

1. bridge to serial buses for consumer electronics: (RC5, D2B)

2. bridge to serial automotive bus: (CAN bus)

3. bridge to parallel buses (Intel/Motorola parallel buses)

## RC5 – I²C Interface Code

### A bit of history

Over the years, a number of remote control systems have come into being. The transmission media have passed from wires (it's true, there used to be wired television remote controls), to ultrasonics and then, to infrared. Coding has, of course, evolved as a function of the problems raised by each of the media and has, progressively, tended towards uniformity.

There are still far too many different remote control coding systems on the market but, with the proliferation of systems using remote control, on the one hand and the number of instructions to be transmitted, on the other hand, there is a tendancy towards consolidation around a few major standards. In Europe, one that is very much employed is the RCx (Remote Control) family. The family has already arrived at its fifth generation. Figure 12.1 reviews the block diagram of the coding/transmission/ reception/decoding present in a remote control device.

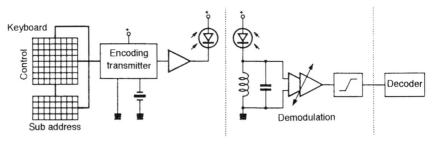

**Figure 12.1**

# The RC5 Code

An RC5 message is serial, asynchronous, time-multiplexed, with clock information embedded in the data. This leads us to refer to a 'message data stream'.

A data stream is nothing but a string of information bits making up a message and, in our case, this data stream represents a binary control code. Obviously, to avoid mixing the codes, it is a good idea to separate the streams from each other by short time intervals to give the receiver time to distinguish one from another. Let's return now to the architecture of the data stream itself. It is made up of bits, words...having precise meanings. Here again not all bits are equal!

## Definition of a message bit

The information represented by a bit can have various electronic manifestations: presence or absence of an electronic level during a defined time, variation of an active duration of a signal during a defined time, change of an electrical level during a defined time, modulation of the position of a signal during a defined time, etc.

Figure 12.2 shows what we have chosen to call a bit in the RC5 protocol. Quite different from many of its brothers, these bits have a rather strange appearance, notably in the position of the transition with respect to the time interval that defines its duration.

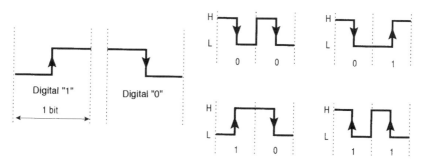

**Figure 12.2**

We can already hear you asking: how do you know that it is a one; we weren't even ready to receive anything...? You are finally going to have the joy of discovering the mysteries of completely asynchronous buses where you are not waiting for anything, you don't know the contents of the message...but which you manage to read nevertheless.

In short, this bit is defined by a rising or falling transition, during (and at the middle of) a defined interval. (Insiders will probably be aware of what is going to happen and can prepare themselves). This two phase system provides good immunity against many types of interference, in particular those coming from various infrared sources (sun, electric lights, open oven doors, radiators...).

## Architecture of an RC5 data stream

We are already in trouble because there are normal and extended RC5 data streams which differ from each other in a few subtle respects. We will first limit ourselves to the conventional RC5 case, knowing that it is used in the greatest number of applications.

The general aspect of an RC5 data stream is given in Figure 12.3(a) and (b). Its format is made up of two distinct parts, but very slightly upstream of the real beginning of the RC5 data stream is an interval of time, corresponding to one or two bits, which is reserved for getting around problems of key bounce. The first zone is a service zone which occupies an interval corresponding to three bits.

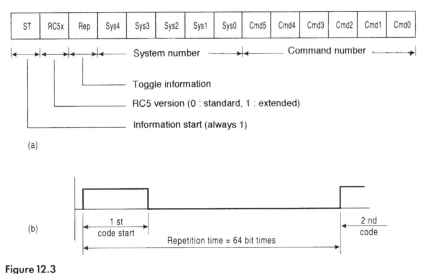

Figure 12.3

## The two first bits

Before developing a software protocol, it is preferable to consult with the hardware designers to have an idea of some of the problems that never fail to come up. Since a remote control is only practical when the user is at a certain physical distance from the receiver, there must be an input stage including an amplifier preceding the decoder. But how can you expect that something inert can guess that you want to communicate with it? It will always be surprised (and, perhaps, disturbed) whenever you try to send it something. To get around this problem, two bits have been installed in the start protocol whose principal purpose is to eliminate the effects of delays and to provide for the establishment of the average value of the automatic gain control voltage (AGC) of the receiver amplifier, regeneration of the clock, synchronisation of the clock and so on. Generally, these first two bits are logic ones.

## The third bit

The function of this control bit is to initialize a new transmission. It can be understood by considering an example: Many keys have repeat or toggle functions for example: the 'sound volume +' key and the 'mute on/off' key. In the first case, repeated pressing of the same key progressively increases the sound level. This key, which mechanically cannot change place between (successive) commands, has no reason to transmit a different code. It is the same for the second case where, after having pressed the key once, the sound is turned off but, pressing exactly the same key a second time (generating the same code) causes the sound to reappear. You now understand that the purpose of this third bit is to indicate either that it is a new transmission or else the retransmission of something already sent.

It would be the same when you select channel 22, since there is no reason for the code to be different for the first and second 'two'. In fact, in this case the exact moment when the key is released the control bit is toggled; that is, it passes from the 0 state to the 1 state and back again each time the remote control is used.

### The following bits

Next follows a zone of directly 'useful' bits which is divided into two fields: five bits of system address which means that the RC5 protocol was designed from the beginning to be able to address $2^5$ or 32 systems (televisions, audio systems, video compact disks, lighting,...), six bits of function control or $2^6$ (64) different instructions per system (louder, fast forward,...), which makes $32 \times 64 = 2048$ control codes!

Anarchy being an integral part of our world, it was soon necessary to react and force users of RC5 to establish a 'standardized' table of 'system/control' codes, rather than letting everyone drown in a sea of haphazardly invented codes. A committee has been set up to establish and distribute and/or attribute new codes as a function of specific individual or corporate needs, whether it be for consumer or industrial applications, demonstrations, etc. Figures 12.4(a) and (b) show the status of codes attributed as of the beginning of 1995.

## Time and frequency

The duration of a bit was chosen as 1.778 ms (about 562 Hz) and, as long as a key is kept depressed, the same message (of $(2 + 1) + 5 + 6 = 14$ bits or 24.89 ms) is retransmitted (entirely) at regular intervals equal to 64 bit periods (113.778 ms), as shown in Figure 12.3(b) Here also, the economic and historical context intervene. These values are the fruit of many considerations. First, the transmitter must consume as little power as possible and, therefore, its internal clock frequency should be as low as possible. Moreover, it should not use costly components – ceramic resonators are recommended.

| Com. | TV1, 2  0,1 | TXT  2,1 | LV  4 | VCR1, 2  5,6 | SATELL  8,10 | CDV |
|---|---|---|---|---|---|---|
| M0 | 0 | | | | | common |
| M1 | 1 | | | | | common |
| M2 | 2 | | | | | common |
| M3 | 3 | | | | | common |
| M4 | 4 | | | | | common |
| M5 | 5 | | | | | common |
| M6 | 6 | | | | | common |
| M7 | 7 | | | | | common |
| M8 | 8 | | | | | common |
| M9 | 9 | | | | | common |
| 10 | 1/2/3 digits | Step page + | Pict. No./time | 1/2/3 dig. /A/P | 1/2/3 digits | Pict. No./time |
| 11 | Freq./prog./ch. | Step page − | Chapter No. | Freq./prog./ch. | Freq./prog./ch. | Recall |
| M12 | Standby | | | | | common |
| M13 | Mute/de-mute | | | | | common |
| M14 | Personal preference | | | | | common |
| M15 | Display | | | | | common |
| M16 | Master volume + | | | | | common |
| M17 | Master volume − | | | | | common |
| M18 | Brightness + | | | | | common |
| M19 | Brightness − | | | | | common |
| M20 | Colour saturation + | | | | | common |
| M21 | Colour saturation − | | | | | common |
| M22 | Master bass + | | | | | common |
| M23 | Master bass − | | | | | common |
| M24 | Master treble + | | | | | common |
| M25 | Master treble − | | | | | common |
| M26 | Master balance right | | | | | common |
| M27 | Master balance left | | | | | common |
| 28 | Pic. contrast + | Enter | TSP once | Today/OTT | | TSP once |
| 29 | Pic. contrast − | Memory out | Repeat cont. | Timer prog. | | Repeat cont. |
| 30 | Search + | Seq. out | Next | Rec. select | | Next |
| 31 | TInt/hue + | | Fast run rev. | Fast run rev. | Preset ant. − | Programming |
| 32 | Ch/prog. + | Exchange | Entry | Step + (up) | Ch/prog. + | |
| 33 | Ch/prog. − | Index/sum. (A) | Auto stop/mem. | Step − (down) | Ch./prog. − | Auto stop/mem. |
| 34 | Altern./ch. | Row zero (A) | Slow run rev. | Slow run rev. | | Slow run rev. |
| 35 | 1/2 language | 1/2 language | Audio 1 | A-o/p select | Lan. sel./mod.1 | A-select |
| 36 | Spatial stereo | Spatial stereo | Audio 2. | Clock/ch./cal. | Spatial stereo | Previous |
| 37 | Stereo/mono | | Speed − | Still rev. | A-mixed/mod. 2 | Still rev. |
| 38 | Sleep timer | Print | Speed + | Speed + | Sleep timer | Speed + |
| 39 | Tint/hue | Antiope/TV (A) | ITR | Speed − | Antennna East | Speed − |
| 40 | RF switch | | Slow run fwd. | Slow run fwd. | TV/sat/key A | Slow run fwd. |
| 41 | Store/vote | Page hold | Still fwd. | Still fwd. | | still fwd. |
| 42 | Ti. pag/ch. clo. | Ti. pag./valid | Fast run fwd. | Fast run fwd. | | dig. mult. scr. |
| 43 | Scan/increm. | L. top/bot./no. | Search auto | Index scan | | Strobe |
| 44 | Decrement | Reveal/conceal | Scan rev. | Scan rev. | | Scan rev. |
| 45 | | Cancel picture | Close/open | Eject | | Open/close |
| 46 | Sec. con/menu | TV/text/sub. (A) | Scan fwd. | Scan fwd. | Subtitle | Scan fwd. |
| 47 | Show clock | | Normal rev. | Normal rev. | Help | Video clip |
| M48 | Pause | | | | | common |
| M49 | Erase (correct or clear entry) | | | | | common |
| M50 | Rewind/fast run reverse | | | | | common |
| M51 | Go to | | | | | common |
| M52 | Wind/fast run forward | | | | | common |
| M53 | Normal run forward (play) | | | | | common |
| M54 | Stop | | | | | common |
| M55 | Recording | | | | | common |
| M56 | External 1 | | | | | common |
| M57 | External 2 | | | | | common |
| 58 | | | Clear mem. A | Short/long play | Antenna West | Clear mem. |
| 59 | Advance | | Freeze segment | Count mem. | Pres. antenna + | A/B program |
| 60 | TXT sub-mode | | TXT/RF switch | TXT sub/SLd | TXT sub-mode | TXT/RF switch |
| M61 | Sys. standby | | | Timer stant | | |
| 62 | Crispener | Newsflash (A) | CX(1, 2, 3) | RF switch | Ant. pos. sta. | CX |
| M63 | System select | | | | | common |

**Figure 12.4**

| 12 | TUNER A 17 | REC 18, 23 | CD 20 | COMEB/PH 21 | SAT. A 22 | P-A 16, 19 |
|---|---|---|---|---|---|---|
| command | | | | | | 0 |
| command | | | | | | 1 |
| command | | | | | | 2 |
| command | | | | | | 3 |
| command | | | | | | 4 |
| command | | | | | | 5 |
| command | | | | | | 6 |
| command | | | | | | 7 |
| command | | | | | | 8 |
| command | | | | | | 9 |
| 1/2/3 digits Freq/prog./ch. | Display scroll | Cursor scroll Display scroll | Display scroll | 1/2 digits Preset/prog | Gr. equal. L Gr. equal. R |
| command | | | | | | Standby |
| command | | | | | | Mute/de-mute |
| command | | | | | | Personal preference |
| command | | | | | | Display |
| command | | | | | | Master volume + |
| command | | | | | | Master volume − |
| command | | | | | | Brightness + |
| command | | | | | | Brightness − |
| command | | | | | | Colour saturation + |
| command | | | | | | Colour saturation − |
| command | | | | | | Master bass + |
| command | | | | | | Master bass − |
| command | | | | | | Master treble + |
| command | | | | | | Master treble − |
| command | | | | | | Master balance right |
| command | | | | | | Master balance left |
| | Repeat once Repeat cont. | TSP once Repeat cont. | TSP once Repeat cont. | | | Gr. eq. L + R Speak sel. |
| Search + Search − | Select + Select − | Select + Select − | Select + Select − | Next sat. Prev. sat. | | High tone fil. Low tone fil. |
| Preset + Preset − | Next Previous | Next Previous | Next/preset + Prev./preset − | Preset + Preset − | | Up Down |
| | Dig. peak, (23) Last rec. cancel Sequential | Index next Index prev. Play/prog. | Index next Index prev. | Prog. + Prog. − | | Sig. path. scr. Speaker A |
| Stereo/mono | Blank search Source/tape | Speed nom. Speed + Speed − | Stereo/mono Sleep timer | Mono A/B | | Surround Sleep timer Speaker B |
| RF switch Store exec. | Rec. pause Store exec. Rec. blank | Store exec. | Synchro start Store exec. | | | Speaker C T-P-S Timer set |
| Scan fwd. | Scan. fwd. Mech. A | Scan fwd. | Scan fwd. | Intro. scan. | | Timer up Timer down |
| FM MW/FM − FM LW | Eject Mech B Normal/rev. dir. | Close/open Fast | Close/open Wave scroll Alarm/buzz | | | Timer mem. Aco. mem. Sel. Aco. mem. |
| command | | | | | | Pause |
| command | | | | | | Erase (correct or clear entry) |
| command | | | | | | Rewind/fast run reverse |
| command | | | | | | Go to |
| command | | | | | | Wind/fast run forward |
| command | | | | | | Normal run forward (play) |
| command | | | | | | Stop |
| command | | | | | | Recording |
| command | | | | | | External 1 |
| command | | | | | | External 2 |
| Clear mem. | Clear mem. A/B program | Clear mem. A/B program Dyn. range ex. | Clear mem. Repeat alarm | | | Clear mem. |
| | | | | | | Dyn. range ex. |
| | Count mem. st | Dyn. ra. comp. | | | | Dyn. ra. comp. |
| command | | | | | | System select |

**Figure 12.4** (Continued)

| Com. | TV1, 2    0,1 | TXT    2,1 | LV    4 | VCR1, 2    5,6 | SATELL    8,10 | CDV |
|---|---|---|---|---|---|---|
| M64 | Surround sound | | | | | common |
| M65 | Balance front | | | | | common |
| M66 | Balance rear | | | | | common |
| M67 | Sound effects | common command | | | Sound effects | Audio int./ext. |
| M68 | Sound effects | common command | | Sound effects | Radio | Video int./ext. |
| M69 | Sound effects | | | | | common |
| 70 | Speech/music | | | A1 picture | | |
| M71 | Diminish brightness | | | | | common |
| M72 | | (Sound or menu function) | | Tracking + | Sound/men. fun. | Fade in |
| M73 | | (Sound or menu function) | | Tracking − | Prev. page | Fade out |
| M74 | | (Sound or menu function) | | NICAM | Next page | Sound/men. fun. |
| M75 | Data str. start | | | Data str. start | | Data stream start |
| M76 | Data str. end | | | Data str. end | | Data stream end |
| M77 | Linear function increment | | | | | common |
| M78 | Linear function decrement | | | | | common |
| M79 | Sound scroll | | | Catalogue | Next prog. | |
| M80 | Step up | | | | | common |
| M81 | Step down | | | | | common |
| M82 | Menu on | | | | | common |
| M83 | Menu off | | | | | common |
| M84 | Display A/V system status | | | | | common |
| M85 | Step left | | | | | common |
| M86 | Step right | | | | | common |
| M87 | (Menu function) | | | Acknowledge | Enter | (Menu function) |
| M88 | PIP on/off | Common video command | | | | PIP on/off |
| M89 | PIP shift | Common video command | | | | PIP shift |
| M90 | PIP/main swap | Common video command | | | | PIP main/swap |
| M91 | Strobe on/off | Common video command | | | | Strobe on/off |
| M92 | Multi strobe | Common video command | | | | Multi strobe |
| M93 | Main frozen | Common video command | | | | Main frozen |
| M94 | 3/9 multi scan | Common video command | | | | 3/9 multi scan |
| M95 | PIP mode select | Common video command | | | | PIP mode select |
| M96 | Mosaic | Common video command | | | | Mosaic |
| M97 | Solarization | Common video command | | | | Solarization |
| M98 | Main picture stored | Common video command | | | | Main picture stored |
| M99 | PIP strobe | Common video command | | | | PIP strobe |
| M100 | Recall main picture | Common video command | | | | Recall main picture |
| M101 | PIP freeze | Common video command | | | | PIP freeze |
| M102 | PIP step up (+) | Common video command | | | | PIP step up (+) |
| M103 | PIP step down (−) | Common video command | | | | PIP step down (−) |
| 104 | PIP size | | | | | |
| 105 | Pic. scroll | | | | | |
| 106 | Act on/off | | | Skip on/off | | |
| 107 | Red (menu) | Red | | Advance | Red | FTS 2 |
| 108 | Green | Green | | 1/2 slow | Green | FTS 1 |
| 109 | Yellow | Yellow | | W TOC/VMI | Yellow | Title |
| 110 | Cyan | Cyan | | Date rec. | Cyan vol. (+) | AMS (+) |
| 111 | White | Index | | Title rec. | Index vol. (−) | AMS(−) |
| 112 | (Next) | | | VISS Index no. | Mail | Index next |
| 113 | (Previous) | | | VISS index pr. | Environ. pay TV | Index previous |
| 114 | | | | Mark | | Next dist. |
| 115 | | | | Erase | | Previous dist. |
| 116 | | | | Re number | Into MAC | |
| 117 | | | | Blank search | | |
| M118 | Sub mode | Common command | | Sub mode | TV 1 | Sub mode |
| M119 | (Options) | (Sub mode) | | | | |
| 120 | | | | Auto repeat | | |
| 121 | | | | ins. ch. select | | |
| 122 | Stone op/close | | | SED ch. mem. | | Store op./close |
| 123 | Connect | | | | | common |
| 124 | Disconnect | | | | | common |
| 125 | | | | C/K set monitor | | |
| 126 | Movie expand | | | Tape select | | |
| 127 | | | | Child lock | Parent access | |

**Figure 12.4** *(Continued)*

| 12 TUNER A 17 | REC 18, 23 | CD 20 | COMEB/PH 21 | SAT. A 22 | P-A 16, 19 |
|---|---|---|---|---|---|
| command | | | | | Surround sound |
| command | | | | | Balance front |
| command | | | | | Balance rear |
| Sound effects | common command | | | | Sound effects |
| Sound effects | common command | | | | Sound effects |
| command | | | | | Sound effects |
| | | | | | |
| command | | | | | Diminish brightness |
| (Sound or menu function) | common command | | | | (Sound or menu function) |
| (Sound or menu function) | common command | | | | (Sound or menu function) |
| (Sound or menu function) | common command | | | | (Sound or menu function) |
| | common command | | | | Data stream start |
| | common command | | | | Data stream end |
| command | | | | | Linear function increment |
| command | | | | | Linear function decrement |
| | | | Sound scroll | | |
| command | | | | | Step up |
| command | | | | | Step down |
| command | | | | | Menu on |
| command | | | | | Menu off |
| command | | | | | Display A/V system status |
| command | | | | | Step left |
| command | | | | | Step right |
| | common command | | | | (Menu function) |
| | | | | | |
| | Aut. start ID<br>Write skip ID<br>Skip on/off<br>Write end ID<br>Write FTS<br>Write TOC<br><br><br>Write start ID<br>Ca/BR/Sk ID<br>Re-format<br>Blank search | Edit<br><br>FTS 2<br>FTS 1<br>Title<br>AMS +<br>AMS − | | 10<br>11<br>12<br>13<br>14<br>15<br>16 | Linear phase |
| | common command | | | | Sub mode |
| common command | | | | | (Sub mode) |
| | | | | Progr. mode<br>Stat. mode<br>Pres. mode | |
| command | | | | | Connect |
| command | | | | | Disconnect |
| | | | | | |

**Figure 12.4** *(Continued)*

Obviously, the duration of a bit is derived from the internal clock frequency of an integrated circuit like, for example, the Philips' SAA3006 and the SAA3010:

$$1.778 \, \text{ms} = 3 \times 2^8 \times \text{Tosc} = 432 \, \text{kHz}$$

In order to carry over the longest distance possible while consuming a minimum of power and remaining insensitive to noise, it is wise to use intelligent signal processing before transmission. (Modulating a carrier directly by the code would expose it to a number of transmission problems.) So this is what we do: after having chosen an infrared carrier with a wavelength on the order of 940 to 950 nm, a sub-carrier is created at the frequency of 36 kHz, which is then modulated by the very slow (562 Hz) RC5 code.

Why all these complications? You will see that the more complicated it is at the sending end, the simpler it will be at the receiving end! Thanks to this procedure, when the 940 nm infrared signal arrives, the receiver only passes a narrow band of around 36 kHz. Obviously, it is important that this band of around 36 kHz be free of major perturbations.

Let us consider several sources of perturbation that are very often present:

+ the sun

+ incandescent and fluorescent lamps

+ IR stereo headphones

+ television and computer monitor time bases...

As an example, Figure 12.5 gives an idea of the spectra of the two types of lamps (operating at a line frequency of 50 Hz), whose harmonics are relatively low in frequency. The problem becomes more complicated when we examine the frequencies of infrared headphones and television time bases.

## Headphones

In principle, IR headphones use the same infrared wave lengths, but with sub-carrier frequencies of 95 to 250 kHz, frequency modulated with deviations of the order of 50 kHz. They should stay out of our way.

## Television time bases

Everyone knows that an American television has a sweep frequency of $525 \times 60/2 = 15\,750 \, \text{Hz}$ and second harmonic at 31 500 Hz. A European

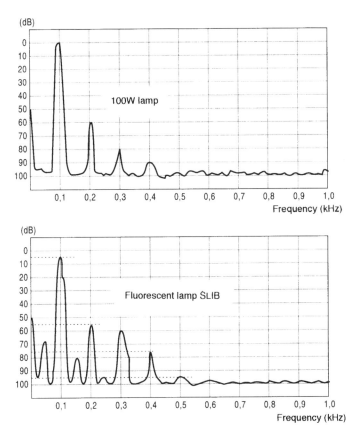

**Figure 12.5**

television has a sweep frequency of 625 × 50/2, for 625 lines, 50 interlaced frames per second, or 15 625 Hz, with a second harmonic at 31 250 Hz. Having adopted a sub-carrier of 36 kHz, we have no particular problems.

Here we are, practically at the end of the fundamental part limited to the generation of the RC5 (non-extended) code.

## Brief resumé of the RC5 code timing (see Figure 12.6)

| | | |
|---|---|---|
| Oscillator frequency: | Fosc | $= 432$ kHz, |
| | F/3 | $= 144$ kHz, |
| | F/12 | $= 36$ kHz |
| Period: | T | $= 27.777\ \mu$s |
| Pulse length: | $tp = 0.25 \times T$ | $= 6.944\ \mu$s |
| Duration of a half bit: | $Tbit/2 = 32 \times 4\ tp = 868.056\ \mu$s |
| Duration of a bit: | $Tbit = Tbit/2 \times 2 = 1.778$ ms |
| Duration of a frame: | $Tfr = 14 \times Tbit = 24.889$ ms |
| Interval between two successive frames: | $T = 64 \times Tbit = 113.778$ ms |

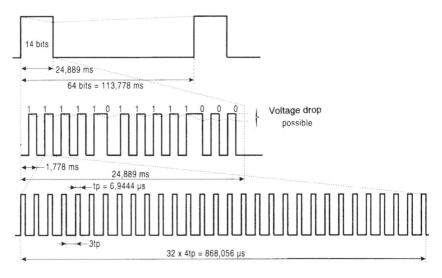

**Figure 12.6**

# Reception of the RC5 Code

The block diagram of the receiver is shown in Figure 12.7. The main performance required of such advice is the following:

+ to amplify correctly the received signals

+ to restore the transmitted RC5 signals correctly so that they can be decoded

+ to be insensitive to powerful transmitted signals

+ to accept rapid variations of the incident signal

+ to be very insensitive to parasitic signals.

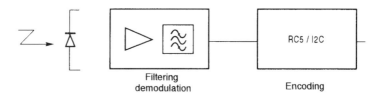

**Figure 12.7**

Many integrated circuits and reception modules of various calibres (according to price...) make reception of the RC5 code easy and reliable.

# RC5 Decoding and the I²C Bridge

We come, finally, to decoding of the RC5 code and to the bridge with the I²C. Figure 12.8 shows the arsenal of solutions proposed. The simplest case: decoding is accomplished by a dedicated circuit (the SAA3049) which presents the entire contents of the RC5 protocol (system, data, etc. . . . ) on easy to use parallel latched outputs, to be decoded (via an HEF 4515) and processed for each application. In this case, the first RC5/I²C bridge consists of placing a PCF8574 I/O port on the output to transfer the information by I²C bus towards a microcontroller.

### A more integrated version

This version includes, simultaneously, the RC5 decoding and its transcoding into I²C. It was realized with a single circuit (SAA3028) and has the interesting feature of being able to simultaneously process a local keyboard also operating in RC5. Although this is fine in itself, with the passage of time, the preferred market solutions have been oriented towards this solution:

Two solutions entirely microcontrolled (using the 80 C51): handling the RC5 by interrupt software: the first solution, entirely software, uses a microcontroller which, itself, has a hardware I²C interface (80 C652, C552, . . . ) which provides bridges to other circuits. The second solution uses an 80 C751 microcontroller (stripped down version of the 80 C51) including a hardware I²C interface, which allows the transformation of this 80 C751 into a more intelligent SAA3028 so that other local tasks can be performed simultaneously. It is this last solution which is the most interesting for designers of systems compatible with the I²C bus.

Since transcoding from RC5 to I²C is not so simple, we propose to examine how that has been done in the case of the second solution. We strongly advise using this approach as a basis for designing software for this kind of bridge. With this approach the circuitry operates on a transition from high to low. The contents of the code transmitted by the RC5 are then transcoded inside the circuit, using its own registers, which can then be reread via the I²C bus. Obviously, the signal needs a little reshaping before the bytes are presented to a microcontroller.

Figure 12.9 shows the output format of the bytes on the I²C bus. As you will notice, this output is made up of 4 bytes and the system number and the control bits are no longer mixed, as they were at the time of the initial transmission of the code.

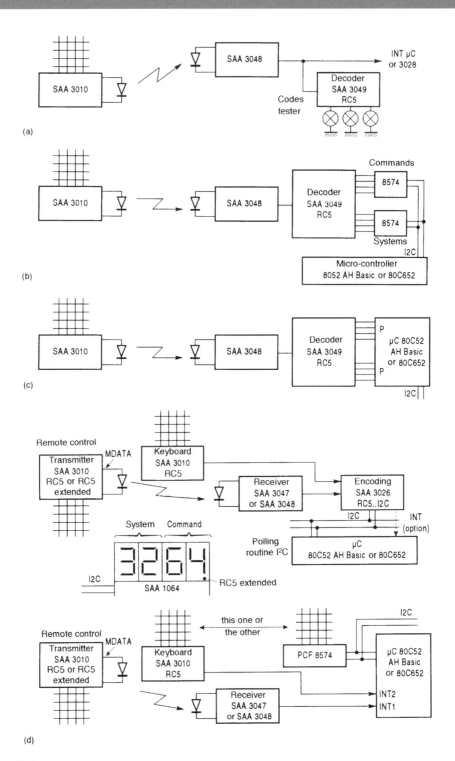

Figure 12.8

Each byte is autonomous and has its own function:

* one containing the system bits

* one containing the control bits

* as you can also see, a byte has been reserved to display the toggle bit, indicating a new message.

Now that everything is in its own cubbyhole, let's examine in detail how that will be handled by the I²C bus.

## Reading the contents of the SAA3028 bytes by the I²C bus

There are two modes for reading data arriving randomly in time:

* polling

* interrupt.

### Polling

In order to determine which technique to choose for these applications, it is worth while to take a look at a particular feature of the circuit operation. We have just seen that, when a single code has been transmitted, recognized as valid, decoded in RC5, and has then (sadly) destroyed its predecessor, it is latched in buffer registers (ready to be read by the I²C).

As of this moment, these registers politely wait until the I²C comes around to read them (and only them). Note, in passing, that it is now up to you to come along. If, by chance, you were momentarily otherwise engaged with more important tasks, you have all the time necessary to read them at your own pace, that is to say, to come back from time to time to scan the SAA3028 circuit to see if, by chance, someone has sent a message (some call this polling, but the word lacks poetic appeal...).

This being the case, after this courtesy call, you will treacherously take advantage of the situation to read the contents of these registers and send them to the microcontroller by taking the bus. If, by chance, you come back too often and, in the interval, nothing new has been transmitted, the registers, which you will have reset during the previous visit, will proudly display a very friendly FF FF FF FF in the bytes that you will bring back....

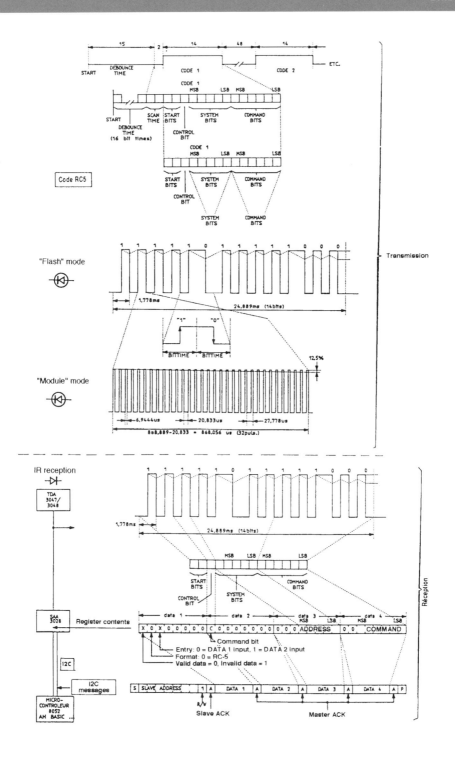

Figure 12.9

On the other hand, you take a considerable risk by using this method. If you do not come back often enough, there is a chance that you will miss messages without having the slightest idea because, we repeat, our book of etiquette says that the following has the right to destroy its predecessor which can be annoying if you happen to be the predecessor! Obviously, you are free to adopt the principle that the last arrival is alway right. In any case, you have been warned!

You should also note that this procedure is easy to put into practice with a microcontroller, because it is only a question of triggering a timer and then going, by the $I^2C$ bus routine, to poll the integrated circuit.

### Interrupt

The SAA3028 circuit ha(s)(d) a pin called DAV (for DAta Valid) which indicates that the received code, after decoding, conforms with the RC5 protocol. This pin can, therefore, be used as an interrupt request for a standard microcontroller, in particular an $I^2C$ microcontroller of the 80 C51 family, such as an 80 C652 ($I^2C$ hardware on the chip) and, in this case, it is via the microcontroller itself that the RC5 – $I^2C$ bridge is realized.

# The D2B (Domestic Digital Bus)

## A short historical review

The D2B bus was designed by PHILIPS early in the 80's mainly for home use and particularly for digital communication between audiovisual equipment At first, this bus seemed a bit ahead of its time and stirred up a good deal of jealousy in other corporate branches (telephone, electrical generation and distribution, building trades,...) coming, as it did, from the powerful and prestigious audiovisual field.

Finally, after considerable effort and partnership agreements between PHILIPS and MATSUSHITA at first then with THOMSON and SONY (representing among themselves 60% of the world market in domestic and audiovisual products), the project arrived at its final phase of standards under the European number EN 1030, later international number CEI (or IEC) 1030, which describes its protocol in detail.

The objective of the D2B is, on the one hand, to interconnect various devices having completely different functions and, on the other hand, to offer to the final consumers (whom we proudly represent) the possibility of not being bound to a particular brand of equipment (Figure 12.10).

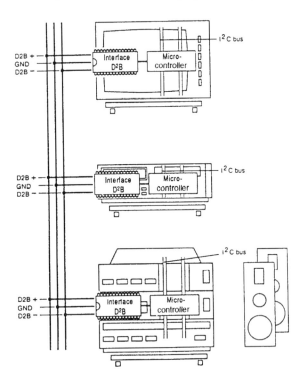

**Figure 12.10**

Obviously the first area of application for this bus, as a result of its initial orientation, is the audiovisual arena. Of course nothing prevents you from using it for other applications, either personal or industrial, because you only have to be compatible with your own codes.

Those of you who are reading these lines with industrial ideas in mind should know that if they would like to use the official D2B logo (Figure 12.11) and have the right to connect to other D2B devices, they should make direct contact with the licence manager, D2B SYSTEMS Co. Ltd. in England.

**Figure 12.11**

## The D2B bus protocol

The objective of this section is not a treatise on the D2B bus similar to that presented for the I²C, but to summarize, in a few words, the make up of a D2B data stream. This bus, operating on a differential pair (with ground return), is asynchronous in nature and the length of its data stream has an upper limit.

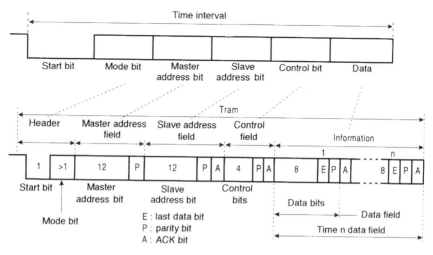

**Figure 12.12**

Figure 12.12 shows the makeup of a D2B data stream.
Examining it attentively, you will find important differences from that of the I²C, mainly with regard to the following points:

  ♦ the electrical definition of a bit is totally different (Figure 12.13)

  ♦ there is a larger address field (master or slave)

  ♦ the master declares its identity

  ♦ arbitration takes place only at the beginning of the data stream

  ♦ there is a continuity bit.

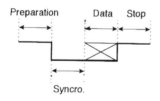

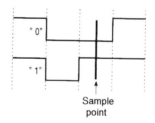

**Figure 12.13**

Before describing how to make a bridge between the D2B and the I²C, and to spare you a long litany on the rôle of each bit, which is beyond the scope of this work, let's take a few lines to quickly summarize the main characteristics.

# Network aspects of the D2B bus

## Type of network

The D2B is a member of the extensive family of LANs (Local Area Networks). For your information, so is its good friend the I²C. The local aspect derives principally from the distance over which information can circulate and not from the quality of the information.

## Length of the network

The D2B network has been designed to operate over a maximum distance of 150 metres. Studies have shown that this distance corresponds to the maximum cable length needed for installation in an individual residence (or apartment), which explains the name domestic (from *domus* which means 'home' in Latin).

## Topology of the network

There are many ways of connecting networks. To satisfy limitations on length, data rate and installation cost, one among them, called daisy chain, has been selected for D2B (Figure 12.14).

# Transmission aspects of D2B

## Symmetry/asymmetry

The I²C bus that we previously presented has two wires (data and clock) which are both asymmetrical with respect to ground. The D2B bus, on the other hand, is symmetrical, using a twisted differential pair with the ground voltage return serving as a shield. This has been done to improve noise immunity from radio frequencies, electromagnetic induction, electrostatics, etc.

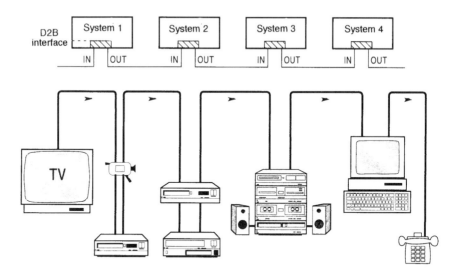

**Figure 12.14**

# The D2B Bus – I²C Bridge

Having finished with the preamble, let's now look at the realization of the I²C D2B bridge (Figure 12.15).

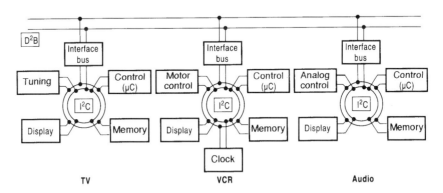

**Figure 12.15**

It is built around the circuit OKI MSM6307 for three reasons:

+ it correctly responds to the standardized protocol (modes 0 and 1) because it was developed under licence, and with the complete agreement of Philips

+ it has been available for several years

♦ it can be interfaced either to the parallel bus of a standard 8 bit micrcontroller or to the I²C bus.

Making use of any or all of the applications previously described for the I²C bus (which is practically limited to a few metres in distance), you can now extend communication over longer distances with the D2B. Figures 12.16 and 12.17 show the block diagram and the pinout of this circuit. (Figure 12.18 shows some typical applications.)

At first observation, the D2B module is ridiculously small! The largest component is the standardized D2B connector! It is true that the digital management of the protocol requires only silicon and, sooner or later, this interface will be integrated on microcontrollers, as has already been done in the case of the I²C bus.

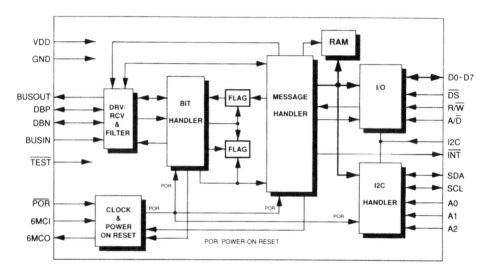

**Figure 12.16**

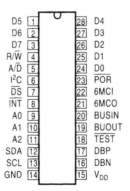

**Figure 12.17**

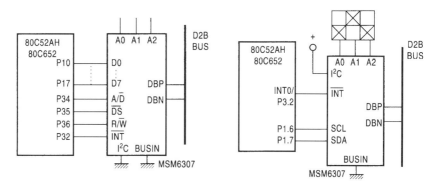

**Figure 12.18**

As you can see, the circuit is mainly designed around a register bank serving as interface between the microcontroller CPU (via the parallel bus) and the protocol manager. The description of the I²C D2B bridge will, therefore, be naturally oriented towards the operation of the MSM6307.

# The MSM6307 registers

The RAM which makes up these 16 registers (from 00h to 0Fh) is divided into four main parts:

* a part reserved for the master
* another part reserved for the slave in receiver mode
* still another reserved for the slave in transmitter mode
* and the last as multi-purpose registers.

Each of these parts is subdivided so as to satisfy the protocol, and has registers of various lengths, adapted to the structure of the D2B bit stream.

The table of Figure 12.19 shows the composition of the registers, their names, their addresses and their operational size. Depending on the operating mode (0 or 1), it is necessary to be able to store a greater or lesser number of bytes of the transmitted bit stream. These registers are accessible, as we have already indicated, either by a parallel bus or by the I²C bus.

Given the many designs in I²C that we have already presented, it is normally easier to load the circuit via the I²C. Obviously, if you are very daring, and you are at the beginning of a specific design, with a microcontroller that does not have an I²C interface on the chip, you can

| HEX | BINARY | ABBR | R/W | #B |
|------|----------|--------|------|------|
| 00 | 00000000 | GL | W | 2 |
| Global register | | | | |
| 01 | 00000001 | LA | R | 2 |
| Lock address register | | | | |
| 02 | 00000010 | MBW | W | 34 |
| **Master buffer (write)** | | | | |
| 03 | 00000011 | MBR | R | 34 |
| **Master buffer (read)** | | | | |
| 04 | 00000100 | | | |
| Not defined | | | | |
| 05 | 00000101 | SRB | R | 34 |
| **Slave receiver buffer** | | | | |
| 06 | 00000110 | STB | W | 16 |
| **Slave transmitter buffer** | | | | |
| 07 | 00000111 | INT | R | 1 |
| Interrupt register | | | | |
| 08 | 00001000 | CLINT | W | 1 |
| Clear interrupt register | | | | |
| 09 | 00001001 | SSR | R | 1 |
| Slave status register | | | | |
| 0A | 00001010 | MCR | W | 1 |
| Master command register | | | | |
| 0B | 00001011 | MSR | R | 1 |
| Master status register | | | | |
| 0C | 00001100 | SRCR | W | 1 |
| Slave receiver command register | | | | |
| 0D | 00001101 | SRSR | R | 1 |
| Slave receiver status register | | | | |
| 0E | 00001110 | STCR | W | 1 |
| Slave transmitter command register | | | | |
| 0F | 00001111 | STSR | R | 1 |
| Slave transmitter status register | | | | |

| Mode | Master-to-Slave | Slave-to-Master |
|------|-----------------|------------------|
| 0 | 2 Data Bytes<br>209 Bytes/sec. | 2 Data Bytes<br>198 Bytes/sec |
| 1 | 32 Data Bytes<br>2457 Bytes/sec | 16 Data Bytes<br>1497 Bytes/sec |

Figure 12.19

always use the parallel bus but, remember it was your idea and you have been sufficiently warned.

## D2B communication, via the I²C interface

Like every standard I²C circuit, the MSM6307 has its own address (in this case 0 1 0 0 A2 A1 A0 = 4xh) which is reconfigurable using external pins. All that remains is to provide it with bytes to transmit on the D2B. Two types of members in the system are capable of sending these bytes:

+ slave-receiver

+ or a slave-transmitter.

Thus we have two transmission formats (via the I²C bus) to initialize the MSM6307 (see Figure 12.20).

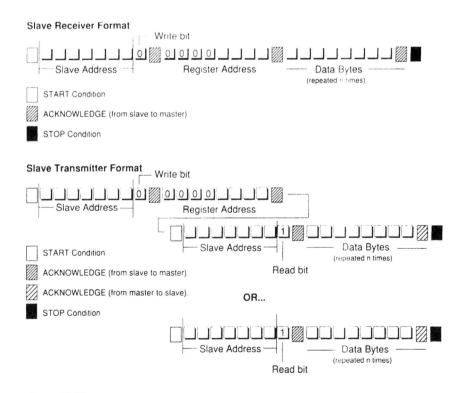

Figure 12.20

In most cases, the D2B interface IC is always in receiver mode in case someone has something to say to it. In this case, using the first format, we can read the contents of the registers via the I²C bus.

In the slave transmitter case, we will use the second format, taking advantage of the clever trick in the I²C protocol which allows reversing the direction of data exchange during the same transmission with the help of a restart condition to change the operation from a Write to a Read.

## Use of the registers

Now that you know the I²C, the D2B protocol and the MSM6307 registers all by heart, you have in your possession the entire arsenal necessary for communication on the D2B bus using an I²C interface.

We will now begin a detailed review of the registers using specific application examples. Let's first start with the simplest case: the MSM6307 as master.

### Master mode

We note that this section includes three registers:

+ a control register

+ a status register

+ buffers.

How do we get all that going? Here is the procedure to use:

### Case of the master-transmitter

1. Load the master's buffer via the I²C (using its real name MBW), at address 02h, which should initially contain the address of the D2B system that you want to address. You should remember that the address of a system connected to the D2B bus is given in 12 bits.

   We are ready to split this 12 bit word into a byte and a half and to place in the first byte the most significant bits of the system address (B11 to B4)

   In the second byte, as shown in the same figure, we will complete the system address and take advantage of the rest of it to add the four bits of control code of the D2B bit stream (see Table 12.1).

**Table 12.1**

| B7 | B6 | B5 | B4 | B3 | B2 | B1 | B0 |
|----|----|----|----|----|----|----|----|
| | | Slave address | | | High and middle nibble | | |
| | | Low nibble | | | Control code | | |
| | | | Data 0 | | | | |
| | | | data n (31 maximum) | | | | |

We then, load the data bytes that we want to transmit by the mile, keeping in mind the operating mode (0 or 1) because, it is only in mode 1 that we have the right to transmit streams 32 bytes long. At this time, all of our little bytes are sleeping quietly in the MBW register, waiting until we want to kick them out.

2. Initialize a master request, using the master control register (MCR = 0Ah) in order to start the transmission. As shown in the following table, the register contains eight bits and we only use the two bits 5 and 7.

| B7 | B6 | B5 | B3 | B2 | B1 | B0 |
|----|----|----|----|----|----|----|
| (1) | mode | X | X | X | X | X |

(1): initialization message

Via the I²C, we simultaneously load the desired mode (0 or 1) in bit 5 and in bit 7. This is the request for sending the D2B data stream to the bus. Knowing that a D2B data stream lasts a maximum of 17 ms, we listen at the microcontroller interrupt pin to find out if, during this lapse of time, the message has been sent.

3. Since you can never be too careful, it is a good idea to read the master status register (MSR at address 0Dh), so as to confirm that the message was sent successfully. Figure 12.21 gives the truth table that shows the quality of the transmission. If everything is just fine, all you need do is:

♦ disable interrupts of the master and clear interrupts

♦ then disable interrupts of the slave and clear interrupts.

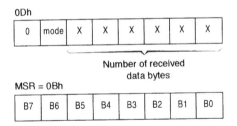

Figure 12.21

Two registers (IR = 07h and CLINT = 08h) give indications as to the status of the interrupts which have taken place or are being processed (see Figure 12.22).

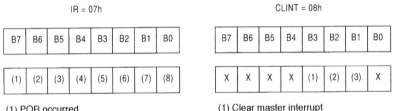

Figure 12.22

The first of these two registers must be tested (by polling, for example) for control purposes and the second must be set in order to be able to function correctly.

Now that the housekeeping has been done, you can continue the transmission (or reception). Before finishing, we have to terminate the master condition.

To summarize, for the master-transmitter case, we give the series of codes that the microcontroller must send via the I²C bus to load the MSM6307 (see Figure 12.23).

Now that you understand the operating mechanism of this circuit, we will now describe the three other operating cases: master-receiver, slave-receiver, and slave transmitter

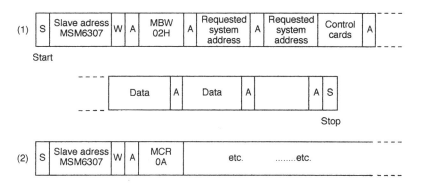

**Figure 12.23**

### Case of the master-receiver

Similar to the case of master-transmitter:

1. Load the master buffer with the address of the slave and the control code

2. Initialize the master request, using the MCR

3. The procedure is established and the reception begins

4. Wait for the interrupt, as before

5. Read the master status register to be sure that everything has gone well

6. Read the contents of the master buffer (=04h) to capture the bytes transmitted

7. Reset the clear interrupts.

It's that simple!

### The slave case

### The slave-receiver case

It is true that this situation is our favourite because we have nothing to do but wait. We have decided on a position of strategic retreat, whereby we listen to the bus and wait until that wonderful day when someone finally remembers that we exist and takes the trouble to talk to us.

In order to talk to us, one has to know our name! It is necessary therefore, at power up, that we decide what our name (address) is (see below the lock address register), and then we need someone to talk to us!

Since D2B messages have been well brought up, they have the good manners, at the end of their bit stream, to lower our letter box flag to let us know that something had been deposited there. What happened, in fact, was that it activated the external interrupt so that our host micro-controller, all excited, could rejoice at having received a letter and hurry to open it as quickly as possible.

Let's be a bit more serious and technical:

1. First, we read the slave receiver status register (in 0Dh) to know how many bytes have arrived

2. Then we read the contents of the slave receiver buffer (in 05h)

3. We reset the interrupts in the clear interrupt register (in 08h)

4. We reset the slave function reception buffers by writing in the slave receiver command (in 0Ch).

For the bit by bit contents of each register, we refer you to Figure 12.24.

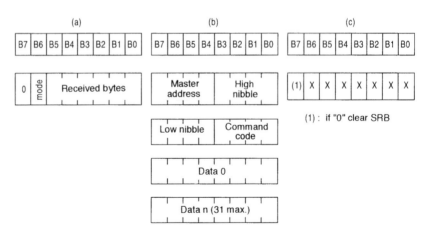

**Figure 12.24**

## Case of the slave transmitter

This is the last possible case!

1. Write the bytes to be transmitted in the specialized slave transmitter buffer, located at 06h (a maximum of 16 bytes)

2. Declare the validity of the data that have just been stored in the buffer with the help of the slave transmitter command (in 0Eh)

3. After the transmission has taken place (when someone has decided to read our message) we will get an interrupt

4. As usual, we take a look at the slave transmitter status, then we clear everything that needs to be cleared.

To help you write the software around this component, it is still necessary to make two or three remarks, of some considerable importance, concerning the registers. We have not yet mentioned the registers called global and lock address register. These two registers encompass several special features of the D2B protocol which we have passed over without comment in the preceding paragraphs, waiting for the appropriate moment to talk about them.

## The global register (GL in 00h)

This register is used to personalize the system at various levels. First, it serves to define the address of the component (and, consequently, that of the system in which it finds itself) using 12 bits (from 000h to FFFh), as shown in the following table.

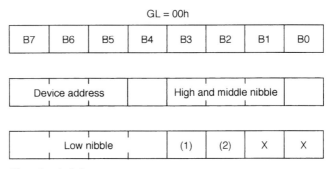

GL = 00h

| B7 | B6 | B5 | B4 | B3 | B2 | B1 | B0 |
|----|----|----|----|----|----|----|----|

| Device address | | High and middle nibble | |
|----|----|----|----|

| Low nibble | | (1) | (2) | X | X |
|----|----|----|----|----|----|

(1): system includes memory
(2): active slave when transmitter

You are, of course, free to assign the address, in your personal configuration, and even to modify it in different time slices, if you wish, but you do that at your own risk.

If you want to assure compatibility with existing, or future, systems it would be well to conform to the codes attributed in the EN 1030 standard.

These codes show that these twelve bits are divided into three groups (4 + 5 + 3).

The first four (most significant) bits of the address indicate the classification of service, and in each of these defined services it is possible, using the five following bits, to designate a particular type of associated equipment. The last three bits can refer to one specific unit among eight possibilities, for a total of 256 different addresses, summarized as follows:

It is up to you to decide on the best use of these addresses.

| From | 0 00h to 0 FFh | communications / telephone |
|------|----------------|----------------------------|
|      | 1 00h to 1 FFh | audio / video / audio-video control |
|      | 2 00h to 2 FFh | house control (kitchen, ...) |
|      | 3 00h to 3 17h | additional screens |
|      | 3 18h to 3 1Fh | additional VCR |
|      | 3 20h to 3 2Fh | additional camera |
|      | 3 30h to F FFh | not defined |

Example:

| (audio / video) | service code | = 0001 | | = 1 | | h |
|-----------------|--------------|--------|---|-----|---|---|
| (video disc) | equipment code | = | 00110 | = | 30 | h |
| (unit n° 4) | unit code | = | 100 | = | 4 | h |
| system address | | = 0001 00110100 | | = 1 | 34 | h |

The last part, or second and last byte of this global register serves also to indicate whether or not this marvelous system, to which the component belongs, has its own memory space (RAM or EEPROM, for example) which could be available as a temporary buffer for storing information. This type of facility can be very helpful when transmitting very long data streams.

## The lock address register (LA in 01h)

There is room for love and passion in the D2B protocol. In a certain fashion, it is possible to declare affection between a requesting and a requested device. This is accomplished through a locking procedure, a sort of: I want you, I will have you, I've got you, game. Obviously, these are only fleeting and/or occasional affairs, and sometimes marriages of convenience.

You will have certainly noticed, in the course of the presentation of the protocol, a table describing the meaning of the four bits contained in the control code part and in which this locked-unlocked condition appears.

It is through this code that a requesting device can lock (or unlock) the request device with which it wants to communicate. You are bound to ask 'what is the good of all that?'. And how right you are! It is not hard too understand. You know that a D2B data stream is limited in length and that, at the beginning of each data stream, an arbitration procedure takes place which takes into account the mode, among other things.

We have explained that priorities exist. Sometimes they can be annoying because, while you are in communication with a certain requested device, it is possible that someone else, using a higher priority, can cut you off because he wants to talk to the same requested device. This poor requested device can very rapidly become confused and fail to understand the messages from the two (or more) requesting devices.

To avoid being perturbed by thoughtless polluters of the verbal environment, the requesting device can lock onto the requested device so as to assure a continued conversation. Since the requested device is polite, it accepts being tied up, ... but only temporarily, because part of the education of the requesting device should be to release its partner when the exchange is complete.

We strongly recommend that you do not forget to unlock the requested device at the end of communication. Your partner may have been taught to remember who it was that tried to force the wedding ring on the unwilling finger and may react discourteously!

After this discourse on the D2B-I²C relations, we are quickly going to bring up another type of I²C bridge used in automobile and industrial applications.

# CAN/VAN Bridges and the I²C

## A bit of history

After the asymmetrical I²C and the differential symmetrical D2B, here are two more children of the numerous LAN family: the CAN and VAN buses.

Deriving its inspiration from the D2B protocol, the German company R. Bosch created several years later, the CAN (Controlled Area Network) bus for industrial use but it was quickly taken over by the automobile sector. Not wanting to be left out, the French groups PSA (PEUGEOT...)

and RENAULT created another protocol of the same type, the VAN (Vehicle Area Network).

Without going into the details of these buses and their architectures, we present the general structure of their data streams in Figures 12.25 for CAN and 12.26 for VAN.

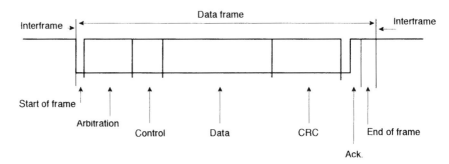

**Figure 12.25**

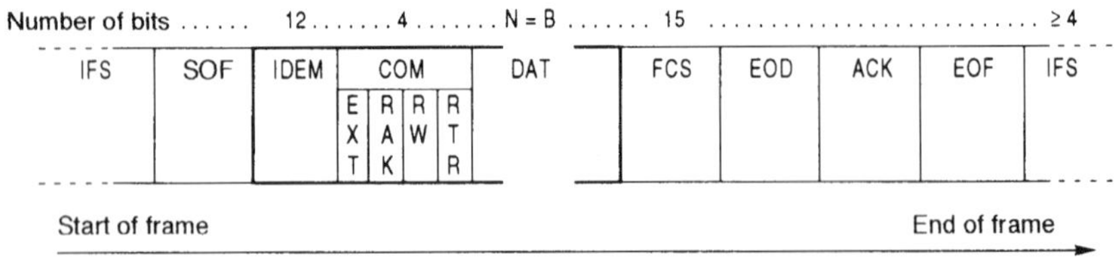

**Figure 12.26**

The family resemblance is evident.

Here also, it is necessary to connect up with microcontrollers, either via their own I²C hardware interfaces or by their 8 bit parallel buses. The latter is the case for both buses, with INTEL or PHILIPS circuits, although I²C interfaces are in preparation for the VAN. For all lovers of the automotive industry, this is something to keep an eye on....

# Part 2

# Interface Between 8 Bit Parallel Buses and the I$^2$C Bus

The main function of the Philips PCF8584 integrated circuit is to form a bridge between a conventional 8 bit parallel bus coming from an Intel/Motorola/Zilog type conventional microcontroller bus, and the I$^2$C bus.

## Operation of the PCF8584

Once more, we have decided to present a circuit in a somewhat unconventional manner. The I$^2$C interface of the PCF8584, although having a slightly lower level of performance than that of the I$^2$C interface on board the 80 C552 or 652, operates just as well in master, slave and multimaster modes.

This circuit design is based on that of the I$^2$C interface of the 84(C) micrcontrollers previously described, to which has been added everything necessary to interface to the 8 bit buses of most of the microcontrollers on the market (8048, 80(C)51, 6800, 68000, Z80, . . .) while maintaining all of the usual facilities – operation by interrupt or by polling, and adding a few special features such as strobe, monitor mode and long distance mode. This allows for connection to the internal buses of all kinds of systems (PC internal bus, for example . . .).

Figure 12.27 and show the pinout and the internal configuration of the circuit. Although we are not in the habit of commenting on circuit pinouts, it is now worth while to analysing them in detail in order to answer certain questions which frequently come up.

### Pin DB0 to DB7

They are simply the parallel data input/output pins which connect to the parallel bus.

### CLK

This is the master clock signal input pin which the circuit needs to create the I$^2$C clock signal SCL, as well as the internal sampling signals necessary

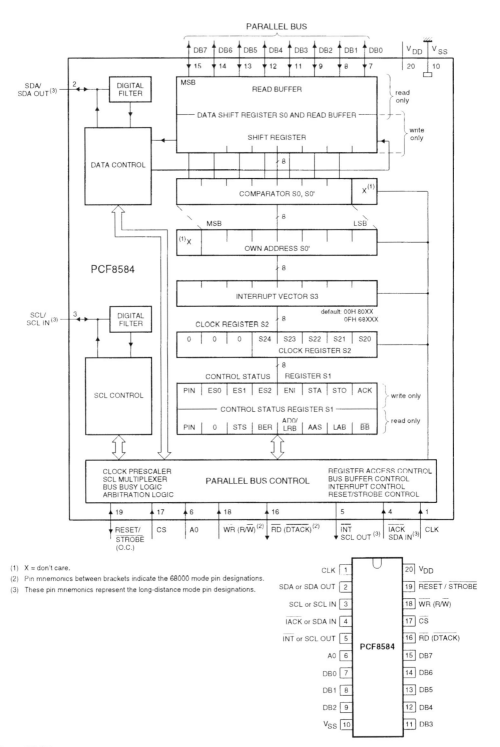

(1) X = don't care.
(2) Pin mnemonics between brackets indicate the 68000 mode pin designations.
(3) These pin mnemonics represent the long-distance mode pin designations.

Figure 12.27

for the digital filtering of the SDA and SCL of the I²C. Since the circuit is not clairvoyant, it is impossible for it to guess the value of the clock signal frequency that you will apply to its pin and takes the default value of 12 MHz for f(CLK). If, for your own reasons, you want to use other frequencies, you must let it know by loading certain specific bits (S22, S23, S24) in one of its internal registers (S2) (see Figure 12.28) so that it can configure its internal dividers to be able to understand your transmissions. Once that is done, the desired value of the I²C bus clock SCL is set via the bits S20 and S21, according to Figure 12.29. We can now feel at ease with all of these clocks.

| INTERNAL CLOCK FREQUENCY | | | |
|---|---|---|---|
| S24 | S23 | S22 | $f_{clk}$ (MHz) |
| 0 | X | X | 3 |
| 1 | 0 | 0 | 4.43 |
| 1 | 0 | 1 | 6 |
| 1 | 1 | 0 | 8 |
| 1 | 1 | 1 | 12 |

**Figure 12.28**

| BIT | | APPROXIMATE SCL FREQUENCY $f_{SCL}$ (kHz) |
|---|---|---|
| S21 | S20 | |
| 0 | 0 | 90 |
| 0 | 1 | 45 |
| 1 | 0 | 11 |
| 1 | 1 | 1.5 |

**Figure 12.29**

Now it gets complicated.

With the exception of CS (Chip Select), which serves to set the circuit in motion, all of the other pins present challenges! To avoid getting stuck, a brief return to the source is useful.

Two major parts can be seen in Figure 12.27. Let's look at the lower part whose purpose is to interface between the various types of parallel buses and the set of internal registers S0, S0', ..., S3. Note that the bus interface is intimately related to the contents of the registers.

How does all that work? Let's begin at the beginning.

At power up (or else after an intentional external signal on the RESET pin), without being told anything else, the circuit thinks that it has been thrown into a system including a microcontroller of the 80 C51 family (default setting).

It is necessary to begin somewhere, and absolutely the first thing that we must do is to give our PCF8584 its name by writing in the registers S0' (Own Address Register). With the best of intentions, the system microcontroller (whose type until now is unknown by the PCF8584) will produce specific signals in order to write in this S0' register. The PCF8584 recognizes these signals, and automatically sets itself to the configuration corresponding to the correct type of micro-controller.

The table below summarizes the various types of signals necessary for, and used by the PCF8584 in the course of these exchanges, depending on the type of microcontroller, thus showing that their recognition can be easily and automatically accomplished.

| Type | $R/\overline{W}$ | $\overline{WR}$ | $\overline{R}$ | DTACK | IACK |
|------|------|------|------|-------|------|
| 8049/51 | N | Y | Y | N | N |
| 68 000 | Y | N | N | Y | Y |
| Z80 | N | Y | Y | N | Y |

# The internal registers and how to access them

## The registers

The PCF8584 contains five registers. Three of them are used for initializing the circuit. Normally, they are written just after circuit RESET. They are:

S0': the I²C address of the PCF8584 (the first to be written!)

S2: the register defining the clocks (already mentioned)

S3: the interrupt vector (see below)
The remaining two others have double functions:

S0: a data buffer register and shift register

S1: a control and status register

In addition, these last two registers can be separately written or read in the course of exchanges (transmission or reception). Let's consider in detail the contents of these registers.

## S0′

Obviously, S0′ (Own Address Register) is only required in the case where the PCF8584 is a slave on the I²C bus. However, for recognition of the microcontroller type, it is recommended to begin by loading S0′ with an address value (a 7 bit value, the most significant bit not being used). Note that, at reset, the internal value of this register is 00 in hexadecimal.

**Remark**: If you load the value ×001 0000 (10 hexadecimal) in S0′, thinking that 10 hexadecimal will be the component I²C slave address, **you will be wrong!**

When an incident message arrives, it enters into S0 and its value (in eight real bits – example: address + R/W̄), in order to be intelligently compared, entirely fills the register whose function is the comparison between S0 and S0′. At that moment, the most significant bit has meaning and represents the highest order bit of the I²C address. In other words, the value written in S0′ has been shifted one step to the left at the level of the comparison register and, although ×001 0000 (10 hexadecimal) was written in S0′, the PCF8584 recognizes itself as ×0010 0000 or 20 hexadecimal, its true address. When the circuit is called and has recognized itself, the bit AAS (Address As Slave) of the Status register S1 is set to 1 for subsequent use.

## S2 clock control register

We have already defined the five least significant bits. The others are not used.

## S3 interrupt vector

This register is designed to contain the value of the interrupt vector so that vectored interrupts can be used. The value of the vector is presented at the parallel port when the interrupt acknowledgement signal is present and the signal ENI (ENable Interrupt) is set. At reset (therefore, in the mode 80 Cxx), its value is 00 hexadecimal and at initialization of the mode 68000, the value is 0F. We can program this register with whatever value we see fit.

## S0 shift register and data buffer

S0 is a combination of a shift register and a buffer register. The parallel data are always written either in the shift register or read from the buffer register. On the other hand, the serial data are shifted (either for input or output) with the help of a shift register. This is the register which is the true border between the serial I²C and our internal parallel 8-bit bus. A little complementary remark here is appropriate concerning data reception: the data contained in the shift register is written into the data buffer only during the I²C acknowledgement phase.

### S1 control/status register

S1 is the most complicated, and we have kept it until last! It is, in fact, two registers, one only for reading and the other only for writing.

*Writing*  the most significant nibble (half byte) serves to route the input or output of information, either directly onto the pins or in the internal registers. It is made up of the bit ESO (Enable Serial Output) whose purpose is to enable (or not enable) the availability of the serial I²C bus, and the bits ES1 and ES2 whose combined function at pin A0 enables access to the contents of the various registers Sx.

In order to limit your suffering, we refer the most fanatic of you to the manufacturer's documentation. For the curious, the tables of Figure 12.30 summarize their functions.

| INTERNAL REGISTER ADDRESSING 2-WIRE MODE | | | | |
|---|---|---|---|---|
| A0 | ES1 | ES2 | I̅A̅C̅K̅ | FUNCTION |
| ESO = 0; serial interface off | | | | |
| 1 | 0 | X | 1 | R/W S1: control |
| 0 | 0 | 0 | 1 | R/W S0': (own address) |
| 0 | 0 | 1 | 1 | R/W S3: (interrupt vector) |
| 0 | 1 | 0 | 1 | R/W S2: (clock register) |
| ESO = 0; serial interface on | | | | |
| 1 | 0 | X | 1 | W S1: control |
| 1 | 0 | X | 1 | R S1; status |
| 0 | 0 | 0 | 1 | R/W S0: (data) |
| 0 | 0 | 1 | 1 | R/W S3: (interrupt vector) |
| X | 0 | X | 0 | R S3: (interrupt vector ACK cycle) |

**Figure 12.30**

The second nibble of S1 (the lower order bits) deal with the management of the interrupt output information, start and stop conditions and acknowledgement of the I²C bus (see Figure 12.31).

*Reading*  In read mode, the register S1 contains, bit by bit, all of the information on the STATUS of the I²C bus. Here also, the manufacturer's specifications are very complete and we will limit ourselves to just a few examples:

| STA | STO | PRESENT MODE | FUNCTION | OPERATION |
|-----|-----|--------------|----------|-----------|
| 1 | 0 | SLV/REC | START | transmit START + address, remain MST/TRM if R/$\overline{W}$ = 0; go to MST/REC if R/$\overline{W}$ = 1 |
| 1 | 0 | MST/TRM | REPEAT START | same as for SLV/REC |
| 0 | 1 | MST/REC; MST/TRM | STOP READ; STOP WRITE | transmit STOP go to SLV/REC mode |
| 1 | 1 | MST | DATA CHAINING | send STOP, START and address after last master frame without STOP sent |
| 0 | 0 | ANY | NOP | no operation |

Figure 12.31

**STS**   this bit indicates the detection of a STOP condition when the component is in the slave-receiver mode,

**BER** ('**Bus Error**')   this bit is set to indicate that a protocol error has been detected on the bus (STOP or START not in their proper places...),

**AAS** ('**Addressed As Slave**')   when the component is in slave-receiver mode, this bit indicates that the circuit has recognized its address on the I²C bus,

**LAB** ('**Lost Arbitration**')   in multimaster mode, this bit indicates that the circuit has lost the arbitration,

**BB** (**Bus Busy**)   indicates that the bus is busy and, therefore, temporarily inaccessible,

**PIN**   this bit is the keystone of the circuit operation. Its management governs the I²C/microcontroller/exchange. As its name Pending Interrupt Not suggests, its purpose is to show whether or not it is time to process the incoming/outgoing I²C data by interrupting the microcontroller (or via polling of the PIN bit).

We will give additional details on this bit when we discuss its control in the software part.

## How to access the registers

Everything that we have said is perfectly true, but one of the problems that we have passed over is how to read or write to all of these registers. We have already surreptitiously slipped in a few words on this subject, but now we will go into it in detail. As you have come to understand, this circuit is composed of several major blocks focused on the most active element: the buffer/shift register S0.

If you have good eyes, you will see that in Figure 12.27, an internal 8 bit bus runs through the entire circuit. It represents the spinal cord of the ensemble through which all interior communication takes place. As for control orders (which register?, Read or Write?) they are obtained by decoding external signals applied to the logic circuits located in the lower part of Figure 12.27. It should also be kept in mind that everything is controlled by the external microcontroller via its 8 bit parallel bus, which is connected to inputs DB0 to DB7, and by the control signals RD/$\overline{\text{WR}}$, AO, INT... and that it is, therefore, under the direct orders of the microcontroller that everything must and does take place.

This being the case, it is enough for the microcontroller to present 8 bits (in parallel) on port DBx and simultaneously (or sometimes slightly before) apply the correct control signals, and the PCF8584 will understand exactly what to do and with which register!

Figure 12.32 summarizes the ways to access these registers.

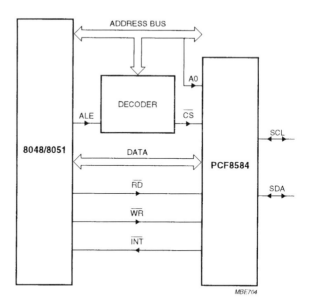

**Figure 12.32**

A picture being worth a thousand words, we will present a little later the complete bone structure of a software example that must be implanted for the circuit to be correctly piloted. We ask a little patience of software lovers.

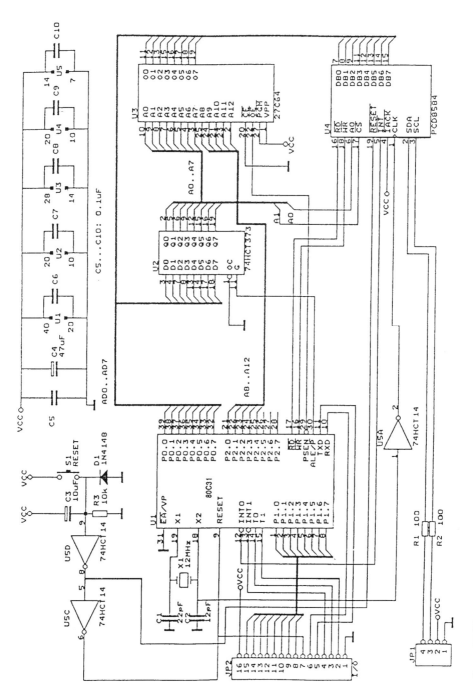

**Figure 12.33**

# Many Applications

## With the 80 C51 family

We now take you back to Figure 12.33 in which an application circuit diagram is shown. No particular technical comments on this subject, but it should be noted that it is often preferable, for reasons of economy, to use the 80 C652, C654 or C552 ... which include the I²C hardware interface on the chip. The present solution is only justified when the entire 80 C52 (Timer 2 included) is used, or in existing 80 C51 solutions that cannot be changed.

## With the 8088 family

Let's remain for a while with INTEL architecture. In this case, the service signals are significantly different and it is necessary to modify the interconnections. This is shown in Figure 12.34.

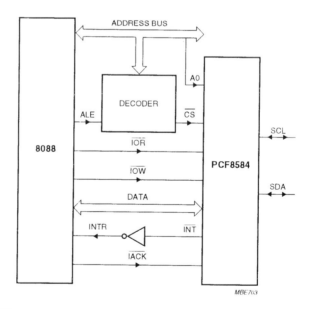

**Figure 12.34**

## With the Z80

Still in the 8 bit family and, although its hour of glory has passed, many of you still use it and should not be deprived of the I²C bus (see Figure 12.35).

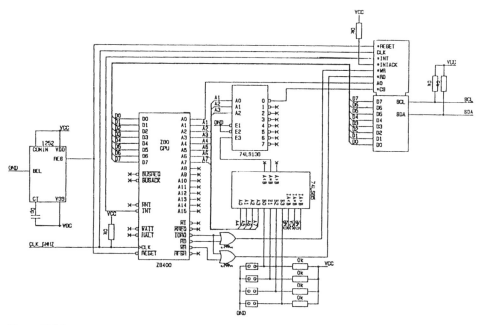

**Figure 12.35**

## With the 68xx and 90 Cxxx family

A change of decor and architecture. This high performance circuit, with a 16/32 bit 68000 at its heart, often needs to deal with us little human beings and, therefore, to use I²C circuits, but its microcontrollers often do not have hardware I²C interfaces. So, long live the PCF8584!

Figure 12.36 gives the general circuit diagram and Figure 12.37 a concrete application of this type of component.

# Applications off the Beaten Track

These are numerous indeed. During the design of the PCF8584, new operating modes were introduced called the Special Mode:

- ◆ a monitor mode

- ◆ a strobe generator mode

- ◆ a long distance mode

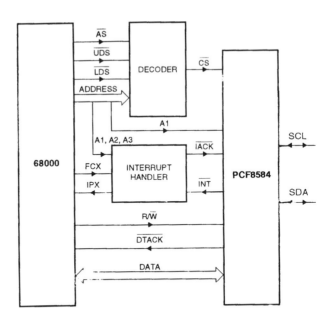

Figure 12.36

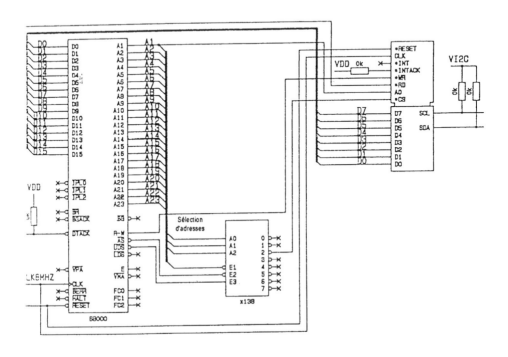

Figure 12.37

### The monitor mode

When the S0′ register is loaded with the value 00 (we have specifically said loaded – an intentional act on your part, because it must be superimposed on the reset value which already is 00 for correct initialization), the PCF8584 has the following special features:

- it is always selected because it listens to everything (for the specialists this is the I²C 'General Call')

- it passes automatically to slave-receiver mode

- it plays dead by not sending any acknowledgement whatsoever

- it goes on strike by not sending any interrupt requests; the data that it receives are directly ready to be read.

In fact, it is a spy. Not very nice..., but very practical, especially if you sometimes want to know what is going on the bus and to develop tools for observing its pranks and flights of fancy. Note in passing, that products based on this device have existed for some time on the market (analysers and I²C bus development aids on PC – see the chapter on Developments Tools).

### The strobe generator mode

The PCF8584 has been initiated into the mysteries of spiritualism! When the circuit receives a message with its own address immediately followed by a STOP condition, it understands that it should indicate its presence by immediately sending a strobe signal (on the active low reset/strobe pin) during eight periods of the clock CLK. This opens up horizons...An example of one of these horizons is shown in Figure 12.38.

### The long distance mode

We have already indicated how to simply and very effectively buffer the I²C bus (Chapter 11) and we have also given some indications on how to use it with opto-couplers and differential pairs. We now show how, with the help of the PCF8584, we can also resolve some of these problems.

First of all, this circuit is provided with a 'Long Distance Mode' which breaks the I²C signals into 4 unidirectional lines, SDA out and SDA in and,

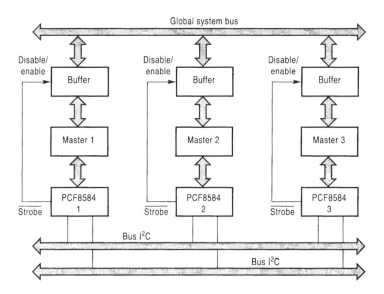

**Figure 12.38**

SCL out and SCL in (see Figure 12.39). These double inputs and outputs have been designed to be connected to simple line driver/receiver circuits controlling long distance lines (example RS232 or RS485 transceivers). See Example 1.

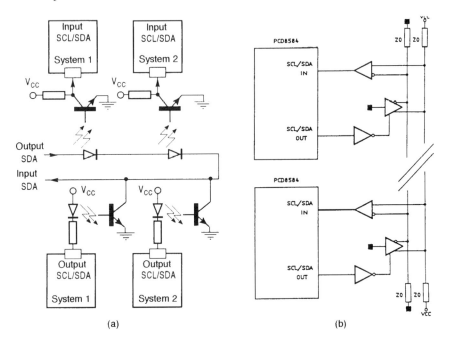

(a)                                                   (b)

**Figure 12.39**

The PCF8584 can also manage the I²C bus using a current loop. To have a current loop, however, you need an input and an output (see Example 2).

It should be noted that in these two types of architecture, the reception of data coming from various circuits placed on the network can only be done (simply) in the polling mode because the interrupt pin is unfortunately already used as SCL OUT in this mode. Two examples of this principle are given in Figure 12.39(a) and (b).

### Example 1

This example shows how to use the inputs and outputs of the same name to transform the I²C bus, whose structure is asymmetrical into symmetrical mode (type RS-485 or RS-422A) using the well known circuits xxx 75176 or XXX 96176, depending on the brand.

### Example 2

This second example offers the possibility of operating in current loop mode while preserving the possibility of being opto-coupled. In this case, when a message is transmitted, the transmission phototransistors being off, the current flows through the diodes into the inputs of the I²C circuits. Once the receiving circuit is interrogated, it has plenty of time to send its message by dynamically pulling the SDA in pin to ground (at the rhythm of the bits transmitted), which allows the master integrated circuit to recover the highly sought-after information.

Using multimaster systems, synchronization and arbitration of the bus are dangerous but, nevertheless, possible.

## Getting the Hardware and Software to Work

The subject being particularly hard to understand, it is worth while returning to the basic operation of the circuit in order to see how it works. Before going once more into the details of its operation, we should remind you succinctly that this circuit, whose purpose is to control the I²C exchange and to manage its protocol, as well as the arbitration procedures and the timing, has a byte oriented data management, conversing with the central processor using either interrupt mode or a polled handshake mode.

The big moment has finally arrived. You have just finished wiring your circuit. Before you do anything else, give it time to catch its breath so that it can discover in what kind of a hornet's nest you have installed it. It is sufficiently grown up to be able to understand automatically, with whom it is intended to communicate.

Complementary to Figure 12.30 already presented, Figures 12.40 and 12.41 show the operating principles of this automatic recognition of the type of bus, which is set in motion by the first WR-CS sequence.

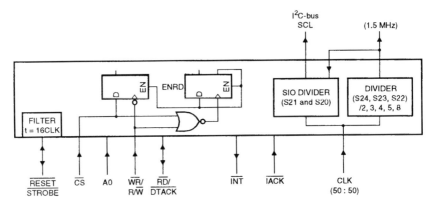

**Figure 12.40**

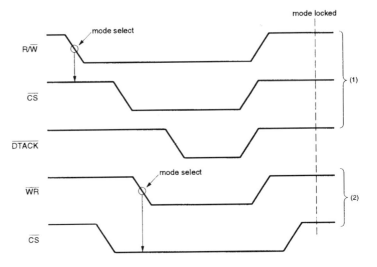

**Figure 12.41**

At this point, one important little remark is in order:

> After the component is powered for the first time, the first WRite signal initializes the PCF8584 (or else in mode 808x, by default).

Now, everything should be going smoothly. Before coming back to our patient software buffs, we want to remind you rapidly of a few basic truths regarding the PCF8584.

## General remarks

The circuit includes five registers (see again Figure 12.27): The first three registers:

S0′: own address register

S2: clock register

S3: interrupt register

are used (mainly) for initialization and are normally written just once after the reset phase of the integrated circuit. The two others which are double registers are accessible both for reading and writing:

S0: register combining the functions of shifting and buffering the data circulating on the I²C bus,

S1: control and status register necessary for accessing and/or monitoring the bus, are used for the transmission and reception of data.

## The PCF8584 control software

The structures of the software examples that we are going to describe in this section are those recommended by the component manufacturer. We haven't changed anything. However, to convince ourselves that there are no bugs in the procedure, we have tested these flow charts with our own software.

### Initialization

The flow chart of an example of an initialization sequence is given in Figure 12.42. The values chosen to illustrate this example are the following:

I²C address of the PCF 8584:  AAh (the value to load, therefore is: 55h)
system clock frequency:  12 MHz
I²C bus data rate or SCL
   frequency:  90 MHz

The flow chart shows precisely the various steps to follow in order to construct appropriate software for initializing the integrated circuit. Note the gymnastics required at pin A0, whose function is to assure the selection of the various registers:

+ when A0 is high the S1 register becomes accessible
+ when A0 is low the other registers can be accessed, according to the values previously loaded in bits ES0, ES1 and ES2 of register S1.

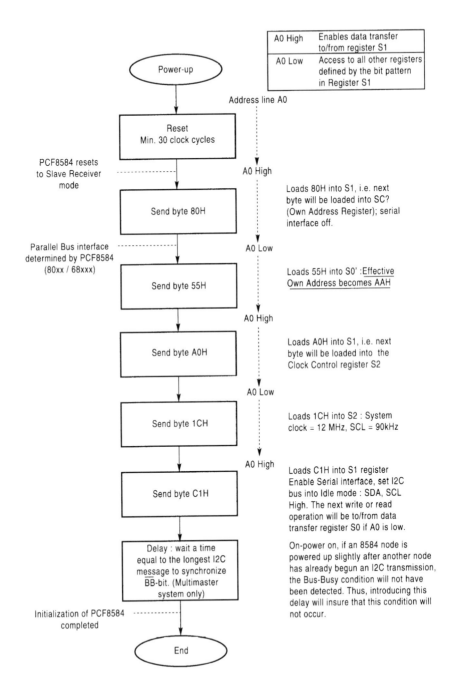

**Figure 12.42**

## The PCF8584 circuit in master mode

Three flow charts summarize the various situations:

Master-transmitter (Figure 12.43)
Master-receiver (Figure 12.44)
Master-transmitter, then receiver after a restart (Figure 12.45).

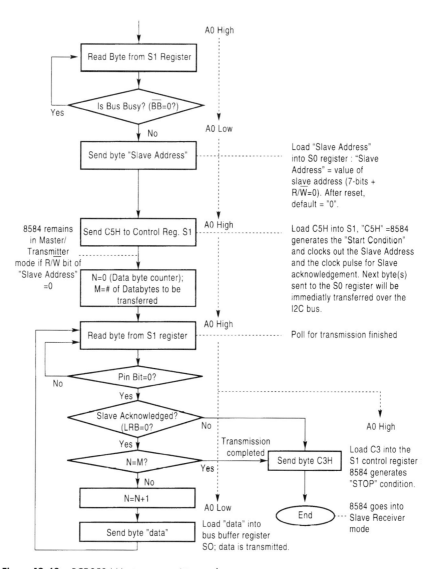

**Figure 12.43** PCF 8584 Master–transmitter mode

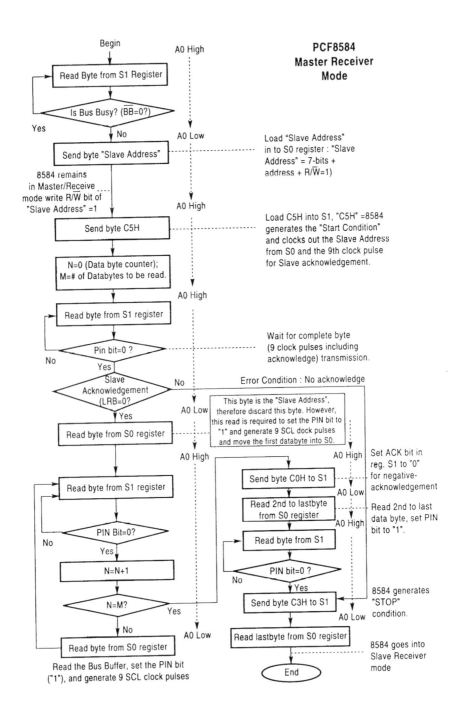

**Figure 12.44**

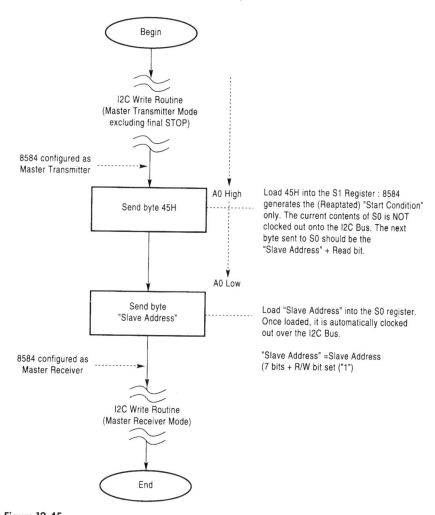

**PCF8584 Master Transmitter-Receiver Mode (Repeated Start)**

I2C Write and Read with "Repeated Start Condition"

Begin

I2C Write Routine
(Master Transmitter Mode
excluding final STOP)

8584 configured as
Master Transmitter

Send byte 45H — A0 High — Load 45H into the S1 Register : 8584 generates the (Reaptated) "Start Condition" only. The current contents of S0 is NOT clocked out onto the I2C Bus. The next byte sent to S0 should be the "Slave Address" + Read bit.

A0 Low

Send byte
"Slave Address" — Load "Slave Address" into the S0 register. Once loaded, it is automatically clocked out over the I2C Bus.

8584 configured as
Master Receiver

"Slave Address" =Slave Address
(7 bits + R/W bit set ("1")

I2C Write Routine
(Master Receiver Mode)

End

**Figure 12.45**

You should take the time to read the flow charts in detail, because they show the entire sequence of operations for single master I²C systems. In addition, for unmitigated hardware enthusiasts (we are proud to consider ourselves in this company) we provide a timing diagram of the signals present on the various pins concerned (Figure 12.46).

Despite all that, it is necessary to add a few comments with respect to the very special register S0.

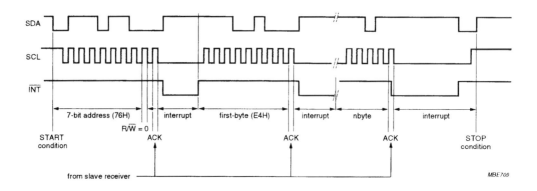

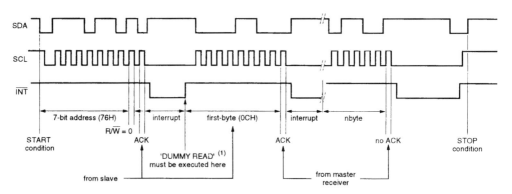

**Figure 12.46**

## S0 register (see Figure 12.47)

The lower part of S0 operates as the shift register for all serial read and write messages on the I²C bus. It can only be accessed by the host microcontroller for writing. The upper part of S0, one of the nicest little parallel registers that you will ever run into, serves as a buffer register for the incident I²C message and can only be read (in parallel) by the host micrcontroller.

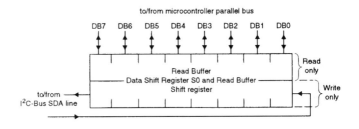

**Figure 12.47**

The questions to ask yourself in order to design good software are:

1. At what time will the data, arriving by the I²C bus in the lower part (shift register), be transferred to the upper part (buffer) so as to be available for reading by the microcontroller?

Answer: during the acknowledgement phase (ninth clock pulse).

2. If the contents of the buffer are not read immediately, what happens?

Answer: so long as the upper part (buffer) has not been read by the external microcontroller, access to the lower part (shift register) for data reception is forbidden and you will be informed of the fact on the pin PIN.

We conclude this section with two remarks.

1. In transmitter mode (whether the circuit is master or slave) the data loaded in the shift register are immediately transmitted to the I²C bus if the enable bit of the serial communication ESO is in the high state.

2. Let's suppose that you want to read the contents of the S0 register (the upper buffer part, of course!) immediately after having written data to be transmitted in the lower part of S0 (actually a common occurrence when we read a byte of information from a slave immediately after having written it's I²C slave address + write bit, sent a subaddress pointer, and then reversed the data direction by issuing a repeated Start + I²C slave address + read bit). It is then necessary to perform a 'dummy read' in order to set up the reception of the correct byte.

Because of the 'double-buffering' of the SO register, a correct reading of incoming data in this case requires reading two bytes: a 'dummy read' to simultaneously clear the value of the buffer and read in our desired byte over the I²C, and the second read to move our precious byte from the shift register to the buffer.

In conclusion, we want to add a few words with respect to Figure 12.46, which shows the flow chart of this case of data transmission immediately followed by reception of data. Contrary to popular opinion, this case occurs frequently, chiefly when using RAM or EEPROM where one is often obliged to write the memory pointer address indicating where reading is to begin, before reading the contents of the memory.

## The slave mode procedures

Figure 12.48 illustrates both slave-receiver and slave-transmitter aspects.

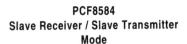

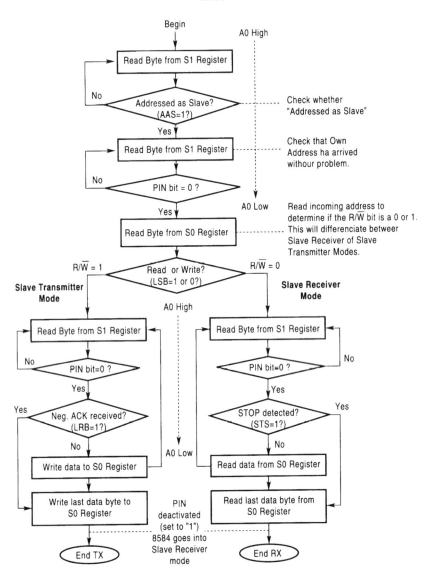

**Figure 12.48**

After the common trunk at the beginning of the routine, the program divides into two independent branches, slave-transmitter or slave-receiver, depending on the value of the last bit (R/W̄) contained in the component's I²C address byte. There are no major complication in this procedure. To provide a better understanding of the circuit operation, we present a timing diagram of the signals appearing on the I²C bus (Figure 12.49).

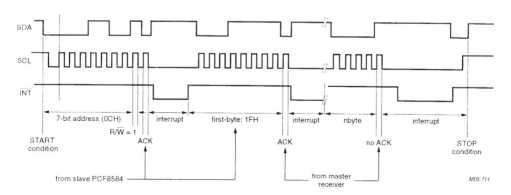

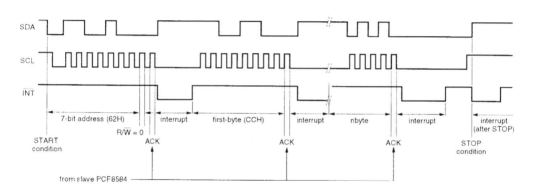

**Figure 12.49**

Normally, the presentation of the PCF8584 and its associated flow charts would end here. It can't help but work correctly... yet, often, this is exactly when trouble begins! In all of your routines you certainly have PIN's here and there and we have shown that this particular bit of the control and status register S1 has a very important place in the overall operation. Here are the details that you have been so much looking forward to.

### The bit PIN (Pending Interrupt NOT) of the S1 register

The PIN bit's main purpose is to inform you as the user of the status of your latest I²C transmission or reception and inform you when you need to make your next move. It is the only bit of it's register that can be both written and read (Figure 12.50).

| BITS | | | | | | | | MODE |
|------|-----|-----|-----|------|-----|-----|-----|--------------|
| PIN | ESO | ES1 | ES2 | ENI | STA | STO | ACK | write only |
| PIN | 0$^{(3)}$ | STS | BER | AD0/ LRB | AAS | LAB | $\overline{BB}$ | read only |

**Figure 12.50**

The PIN bit is a status flag which is used to synchronize serial communication over the I²C bus and is set to 0 (active) whenever the PCF8584 requires servicing. The PIN bit is normally read in polled applications to determine when an I²C-bus byte transmission/reception has been completed. Each time a serial data transmission is initiated (by setting the STA bit in the same register), the PIN bit will be set to 1 automatically (inactive). When acting as a transmitter, PIN is also set to logic 1 (inactive), each time the data register S0 is read.

After transmission or reception of 1 byte on the I²C-bus (9 clocks, including acknowledge), the PIN bit will be automatically reset to 0 (active) indicating a complete byte transmission or reception. When the PIN bit is subsequently set to 1 (inactive), all status bits will be reset to 0. PIN is also set to 0 on a BER (bus error) condition.

In polled applications, the PIN bit is tested to determine when a serial transmission/reception has been completed. When the ENI bit is also set to 1, the hardware interrupt is enabled. In this case, the PIN flag also triggers the external interrupt (active low) via the INT output each time PIN is reset to 0.

When acting as slave transmitter or slave receiver, while PIN = 0, the PCF8584 will suspend I²C-bus transmission by holding the SCL line LOW until the PIN bit is set to 1. This prevents further data from being transmitted or received until the current data byte in S0 has been read (when acting as slave-receiver) or the next data byte is written to S0 (when acting as slave-transmitter). This prevents us from losing precious bytes into oblivion!

We have now come to the end of this long chapter on bridges to the I²C which will allow you to communicate more easily with other systems.

# 13 I²C Bus Evaluation and Development Tools

## Introduction

In this chapter, we inventory the tools that make it possible to simulate, evaluate and emulate the I²C bus to optimally fit our applications and budgets. Over the years many different manufacturers have introduced I²C demonstration boards and I²C control and monitoring equipment ranging in price from a few hundred dollars to thousands. New products are being introduced all the time: this chapter merely gives you an idea of what is available at the time this book was written.

These tools address these basic 4 areas:

- What to do to simulate the bus
- How to evaluate the operation of an I²C integrated circuit
- How to spy on the bus to find out what is really going on
- How to develope, emulate and debug I²C software.

Many solutions, from very simple and cheap to complex and expensive, have been developed to address these issues. Let's begin by describing the simplest and most trivial solution.

## The Poor Man's I²C Generator

We're not going to spend much time on this solution, however it can really come in handy sometimes. Figure 13.1 shows how to pilot the I²C bus with two debounced switches, two LED's and lots of patience and courage.

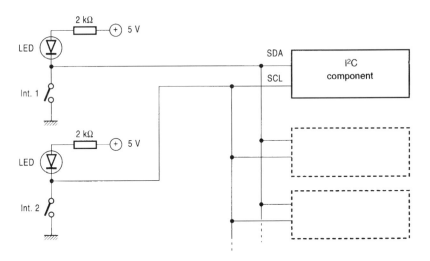

**Figure 13.1**

This is not the best solution in terms of speed, flexibility or user-friendliness, however it the cheapest, and actually works because the I²C is static and can be run as slow as a turtle. Any single-master I²C system can be controlled with this method!

# The Philips Semiconductors Official Version

PHILIPS offers the OM1022 I²C Controller kit (multimaster) and OM4777 I²C Controller kit (single master) that simulate the bus by using a PC together with a printer-port adapter card and special menu-driven software.

This system consists of:

♦ a floppy disk with menu-driven software modules that control many I²C peripherals via dedicated menus, plus a universal I²C bit stream generator for generating your own 'tailor made' messages (see Figure 13.2)

♦ printed circuits, either single master or multimaster, whose circuit diagrams are given in Figure 13.3. Both connect to the PC parallel port.

These kits allow your PC to control and monitor an I²C bus, control specific peripheral devices, or create your own tailored I²C messages. The interface software is menu-driven and works best under a DOS environment.

These kits also allow the development, testing and debugging of I²C hardware. In our opinion, it is an indispensable, simple yet high-performance tool for developing and testing a prototype circuit.

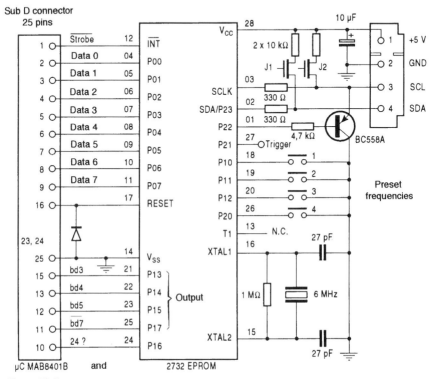

**Figure 13.2**

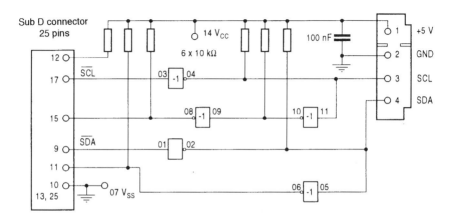

**Figure 13.3**

To satisfy manufacturers using the I²C bus in mass production (production lines, control and test stations, etc.), several companies have created PC add-on cards that operate on compatible PC's, enabling users to monitor and control the I²C bus, and/or develop software in cross assembler, C, or Pascal that dynamically controls the I²C bus. Here are some examples; the first two developed around the PCF8584:

## Calibre ICA-90

This is a plug-on, half length IBM-PC compatible I²C adapter card. This PC card interfaces to the I²C bus via a 9-pin D connector. This card is based on the Philips PCF8584 I²C bus controller IC, which is interfaced to the PC bus. The 8584 can interface to the bus at high speed. Calibre supplies the board with a library of I²C control routines in both C and Turbo BASIC, which can be called by the user's application software. These routines support both master and slave operation. The software is not interactive, i.e. the user must write and compile his own programs, but the interface to the library routines is straightforward and examples are supplied.

A stand-alone monitor program is supplied with the ICA-90. It allows non-intrusive, real time tracing of I²C bus activity. The data captured is stored in PC memory, and when the buffer is filled, trace stops and the data is formatted and moved into a disk file. Data presentation includes occurrences of Start, Stop and Acknowledge conditions. The user can display and analyse the data with any word processor or browsing program. This card provides real-time I²C-trace, a capability that can be very useful to anyone debugging a complicated and or multi-master system.

The monitor program requires at least a 6 MHz 286 based PC or faster. This board is recommended for speed-critical and or complex I²C systems (i.e. multimaster) due to it's real-time monitor capabiltiy.

Calibre UK Ltd. is based in Bradford, England, tel (+44) (0)274-394125. Their representative in the USA is Saelig Company.

## I²C/ACCESS.bus monitor MIIC-101

The MIIC-101 is a stand-alone trouble shooting tool for the I²C and ACCESS.bus. When connected to an I²C bus or ACCESS.bus network, the 101 Bus Monitor can collect, display or upload information on all bus activity.

## Key features

+ $I^2C$ and ACCESS.bus compatible

+ Operating modes: Lines status, Forward/backward trace, view and remote

+ Monitoring of all or selected bus addresses

+ Trace Buffer stores up to 2700 messages

  Easy to read alphanumeric display. Byte, message and buffer scrolling

+ Hand-held portable unit. Battery, external supply or Bus-powered

+ RS232 Port supports remote data capture and uploading

Manufactured by:
> **Micro Computer Control Corporation**
> P.O. Box 275
> 17 Model Ave.
> Hopewell, New Jersey 08525 USA
> tel (+001) 609 466-1751
> fax (+001) 609 466-4116

## FLUKE; PF8681 $I^2C$-bus and ACCESS.bus analysis support package

The PF8681 has been designed for use with the PM3580 family of logic analysers. It provides facilities for analysing and troubleshooting data streams on the $I^2C$ and ACCESS.bus.

Captured data from either bus can be displayed on the logic analyser screen in various number systems. The PF8681 includes a disassembler for the $I^2C$ bus and for the ACCESS.bus. The adapter allows simultaneous measurements in the timing and state domain without any reconnection or multiple probing of the $I^2C$ signal lines. This single probing approach avoids additional DC and AC loading of the $I^2C$ and ACCESS.bus signal lines. The $I^2C$ bus disassembler supports all present day features of the $I^2C$ bus system including 10-bit and 'fast mode'. The ACCESS.bus disassambler supports the BASE-protocol specification as mentioned in the ACCESS.bus specifications version 2.0.

The $I^2C$/ACCESS.bus package PF8681 includes the adapter, disassembler and special ACCESS.bus interface cable. Pricing and delivery information is available from FLUKE.

These last two cards, delivered with floppy disks and very clear and complete documentation, are professional tools whose performance is excellent, within the operational scope of each.

# Simulators and Emulators of I²C Microcontrollers

Many companies supply emulators for microcontrollers with I²C hardware interfaces. A partial list includes the derivatives of the 80 C51 family, such as the 80 C751, 652, 654, 535, 552, 528. You can rest assured that any microcontroller with integrated I²C interface has a corresponding PC-based in-circuit emulator! Contact your nearest Philips representative for the latest tools.

Most of these companies, such as Nohau, Ashling, Raisonance and others, have called upon the microcontroller manufacturers to supply them with 'bond out' versions (the actual microcontroller with program memory address and data bus lines pinned externally) to produce their emulation probes, enabling them to provide 100% emulation of the target micro, and provide 100% realtime I²C emulation as well. Some of these products can even simulate the I²C slave responses, making the emulator simulation even more flexible.

## I²C evaluation and demonstration board

The Philips OM5027 evaluation board provides a working I²C bus system which may be used for familiarization with I²C bus hardware and software. The board features the Philips 80 CL580 (or 83 CL580) microcontroller, and provision for a 40 pin 8051 derivative or external emulator such as the PDS51 Development System. In addition to the central I²C bus there is provision on the Board to connect either of the microcontrollers to an RS232 interface circuit. This PCB also features eleven I²C peripheral circuits which may be switched onto a central bus as required via DIP switches. The modular construction and central bus allow the board to be used as a quick prototyping aid as well as providing for extension to other circuits as required.

The board is supplied with demonstration software programmed into the EPROM.

The Evaluation Board requires power from a 7 to 12 V plug pack or similar DC power supply (50 mA max.) and consists of the following modules:

+ Microcontroller 80 C51-based 80 CL580 (or 83 CL580) with software in 27 C256 EPROM

- Socket for emulator control compatible with 8051 derivative or external development system

- LCD (segment display)   Philips LPH3802-1 (7 segment, starburst, etc.) with LCD driver IC

- PCF8578 mounted on the glass (Chip-on-glass module)

- LCD (dot-matrix display)   Philips LPH3827-1 (3 line × 10 characters, 5 × 7 dot matrix) with LCD driver IC PCF2116A mounted on the glass (Chip-on-glass module)

- I²C to parallel I/O   PCF8574P, 8-bit remote I/O port with edge connector

- I²C to parallel I/O   PCF8574P, 8-bit remote I/O port with alternate I²C slave address used in a 8 multiplexed pushbutton/2 LED drive application example

- RAM   PCF8570P low-voltage 256 byte static RAM

- EEPROM   PCA8581CP 128 byte low-voltage EEPROM

- Clock/Calendar   PCF8593 low-power Clock/Calendar with crystal and battery backup

- DTMF/Tone generator   PCD3312P with TDA1070 audio amp, and piezo sounder

- A/D, D/A   PCF8591 8-bit, 4 channel multiplexed A/D, plus 1 channel D/A

- Parallel bus to interface I²C   PCF8584P Motorola/Intel parallel bus to I²C bus interface IC

- I²C extender/booster   82B715 I²C extender/booster allowing 10x bus capacity

- Power supply board   Plug pack socket, variable 2.5 to 5.0 V voltage regulator, LED indicator

- RS232 interface   MAX232 and 9 Pin RS32 D Connector. RS232 cable included

♦ Software   Documented Demo source code on disk

♦ User's Manual   Hardware description, demo instruction, and schematics.

The OM5027 is available from Philips Semiconductors.

SECTION **5**

# APPENDICES

Appendix 1

Appendix 2

# 1

# Appendix

The printed circuits and their layouts

+ I²C CPU to 8x C552 card

+ CPU I²C to 8x C51/52

+ 8x C652/654 and 8052 AH Basic card...

+ Parts list for the CPU card

+ CPU card layout

+ I²C modules

## CPU part list

| Resistors 1/4 W, 5% | | Capacitors | |
|---|---|---|---|
| R1, R4 | : 820 Ω | C1 | : 470 μF/25 V |
| R2, R3, R9 | : 330 Ω | C2, C3, C4, C9, C15 | : 100 nF polyester |
| R5 | : 56 kΩ | C5 | : 4.7 μF/63 V |
| R6 | : 100 Ω | C6 | : 3.3 nF ceramic |
| R7 | : 10 kΩ | C7 | : 2/22 pF Philips |
| R8 | : 22 Ω | C8, C10 | : 22 pF ceramic |
| R10 | : 68.1 kΩ 1% | C11, C12, C13, C14 | : 10 μF/63 V |
| R11 | : 9.09 kΩ 1% | | |
| P1, P2 | : Potential metre T9YA 1 kΩ 1/4 W, 10% | | |

| IC's | | Misc. | |
|---|---|---|---|
| (U) IC1 | : 7805 | Q1 | : quartz 12 MHz HC 18 |
| IC2 | : 78 L 05 AC | Q2 | : quartz 32.768 kHz MX 38 |
| IC3 | : PCB 80 C 552-4 WP PHILIPS | B1 | : batteries 3.6 V |
| | | | 100 mA (M 36 Aglo) |
| IC4 | : 74 HCT 573 | PF1 | : fuse for circuit (5 × 20) |
| IC5 | : 27 C 512 (EPROM 64 k × 8) | F1 | : fuse 5 × 20 200 mA |

# CPU part list (continued)

| **IC's** | | **Misc.** | |
|---|---|---|---|
| IC6 | : 62256 (RAM static 256 k × 8) | RD1 | : heatsink WA 400 Schaffner |
| IC7 | : PCF 8583 | SC1 | : PLCC 68. Augat sockets |
| IC8 | : PCF 8582 A | SC2 | : DIP 28. Augat sockets |
| IC9 | : PCF 125-0 (supervisor of | J1, at J6, J9 | : HE 14 plug |
|  | of tension Philips) | K1, at K3 | : jumper SANTEC |
| IC10 | : LT 1080 (RS 232 CMOS) | S1, at S3 | : 3 points. Sockets |
| IC11 | : 74 HCT 08 | J7, J8 | : connection |

**Semiconductors**

| | |
|---|---|
| D1, D3, D4, D5 | : BAT 85 (schottky) |
| D2 | : 1N 4007 |

| **Capacitors** | | **IC's** | |
|---|---|---|---|
| C | : 100 nF | IC1 | : 8052 AH BASIC |
| C1 | : 22 pF | IC2 | : 74 HC 08 |
| C | : 22 pF | IC3 | : 74 HC 138 |
| C3 | : 10 μF/10 V | IC4 | : 74 HC 573 |

| **Resistor** | | **Misc.** | |
|---|---|---|---|
| RS1 | : SIL 8 × 10 kΩ | Q | : 11.0592 MHz |
| R1 | : 330 Ω | RAM | : 8 K × 8 (5564 . . .) |
| R2 | : 33 Ω | EPROM | : 8 K × 8 (27C64 . . .) |
| R3 | : 1.5 kΩ | D | : 1N4148 |
| R4 | : 1.5 kΩ | | |
| R5 | : 10 kΩ | | |
| R6 | : 8.2 kΩ | | |
| R7 | : 10 kΩ | | |
| R8 | : 1 kΩ | | |

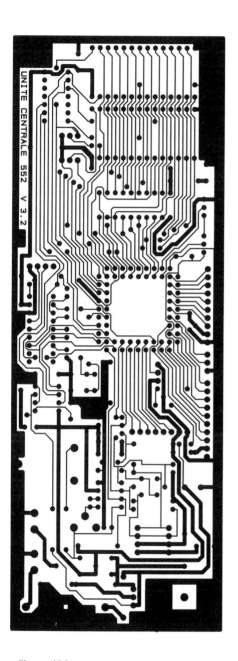
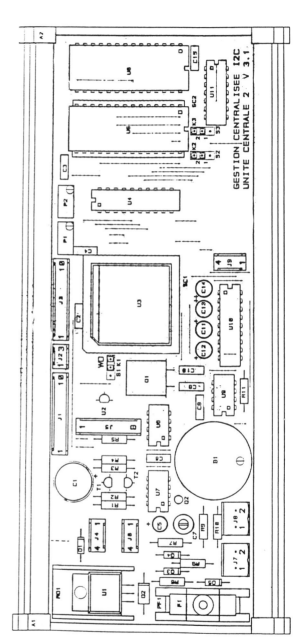

Figure A1.1

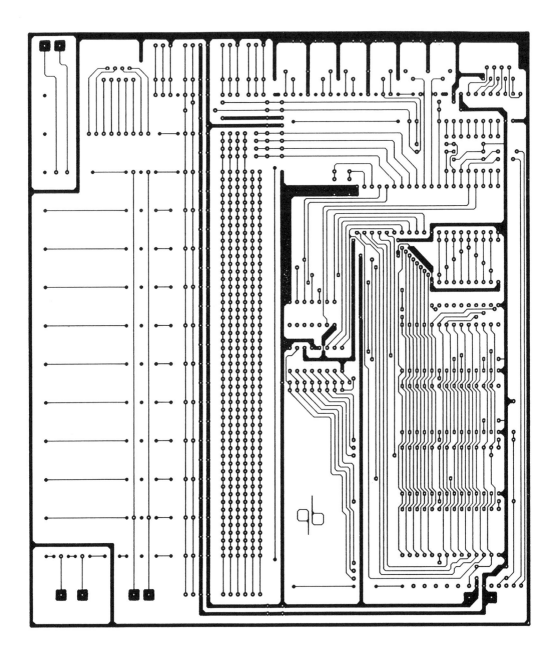

Figure A1.2

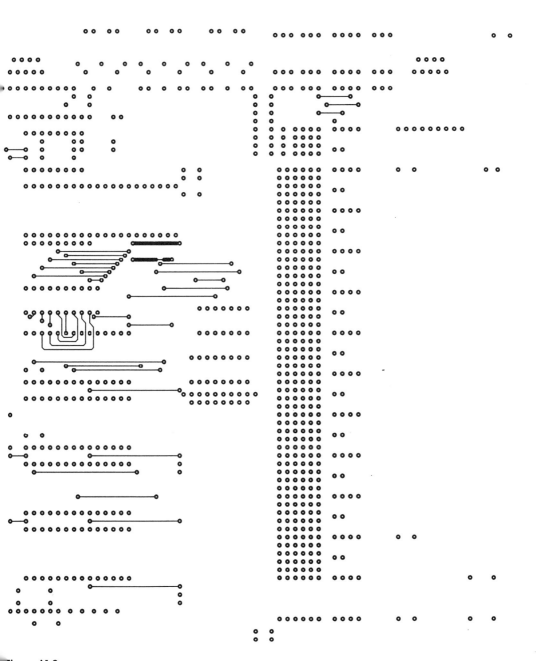

Figure A1.3

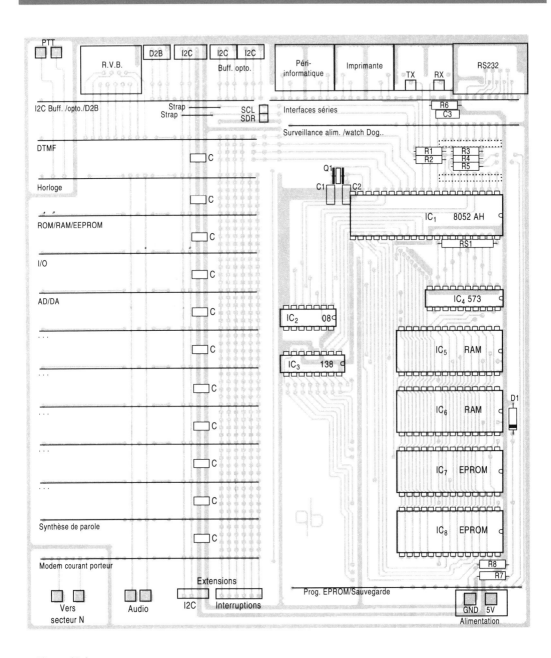

**Figure A1.4**

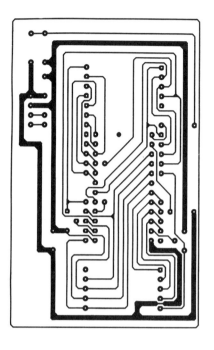
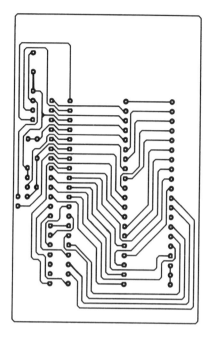
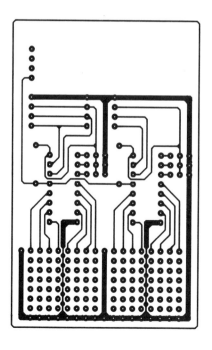
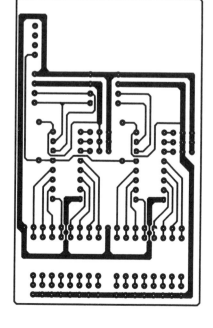

Figure A1.5

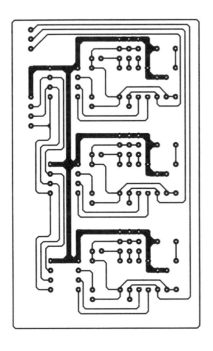
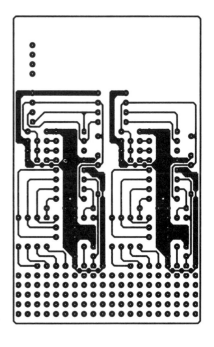

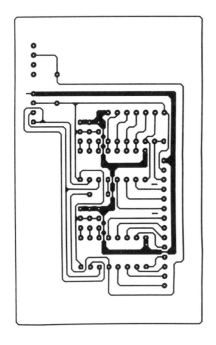
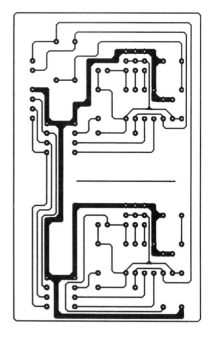

Figure A1.6

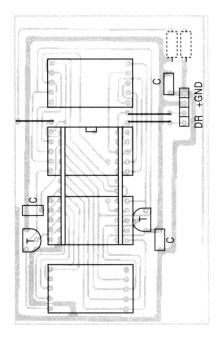

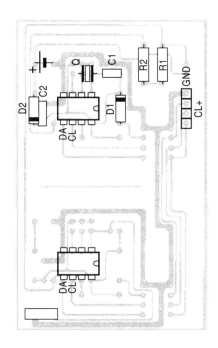

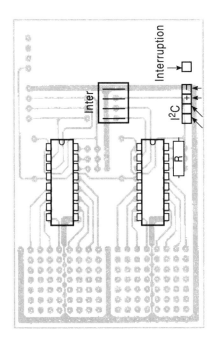

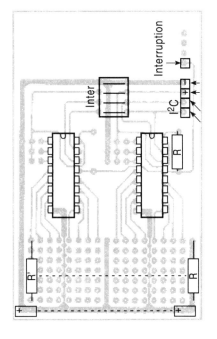

Figure A1.7

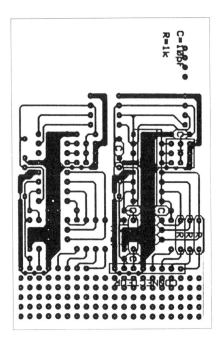

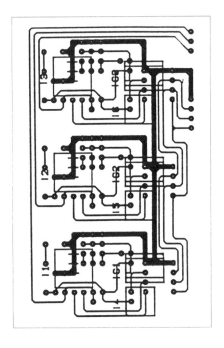

Figure A1.8

# 2 Appendix: Software Examples for the I²C bus

The floppy disk supplied contains a number of examples of software whose main purpose is to serve as a framework for your own software. These examples have been designed for specific hardware architectures which are not necessarily the same as yours. Although their structures are quite general, you will certainly be obliged to modify them. For this reason, they have been well documented, either in the various chapters of this work or, directly at the level of the 'source' files.

You will find on this floppy disk:

+ I²C control logic for standard 80C 51/52 or for the 8052 AH BASIC

+ Control software for the I²C hardware interface of the 80 C652 or C552, written in assembly language and in 'C'

+ Control routines for the principal I²C modules described in this work (in assembly language)

For the most part, we have attached their assembled versions, up to the level of the hexadecimal codes (in Intel hexa i8h), ready to be loaded in (E)PROMS.

In conclusion, it would be thoughtless not to thank again several friends who were kind enough to participate in the preparation of much of this work. Once more, then, many thanks to Mesdames Marie Laurence CIBOT-DEVAUX and Blandine DELABRE-GARNIER and Messrs. Jean Pierre BILLIARD and Olivier SALÉ.

We wish you well in your work and hope that you will have a good time.

See you soon, perhaps, for another bus...

Dominique PARET

# Index

*Index compiled by Geoffrey Jones*